Interstellar dust grain: diameter 4×10^{-6} inches

Blue light wavelength: 1.9×10^{-5} inches

Bacterium: diameter 4×10^{-5} inches

Black hole: diameter 40 miles

Large moon crater: diameter 120 miles

Largest asteroid: diameter 456 miles

Mars: diameter 4,078 miles

White dwarf: diameter 6,000 miles

Venus: diameter 7,251 miles

Bright young stars blaze in the spiral arms of the celestial pinwheel NGC 2997.

Ten million light-years from Earth, vast dust clouds dapple the galaxy NGC 253.

The elliptical galaxy NGC 5128 is banded by what may be the remains of another star system.

The Great Galaxy in Andromeda, serene sister to the Milky Way, dwarfs two satellites.

The barred spiral galaxy designated NGC 1365 flings luminous arms from a central bar.

Seen edge on, a dust lane defines the central plane of the glowing Sombrero galaxy.

GALAXIES

TIME
LIFE ®

This volume is one of a series that
examines the universe in all its aspects,
from its beginnings in the Big Bang to the
promise of space exploration.

VOYAGE THROUGH THE UNIVERSE

GALAXIES

BY THE EDITORS OF TIME-LIFE BOOKS
ALEXANDRIA, VIRGINIA

CONTENTS

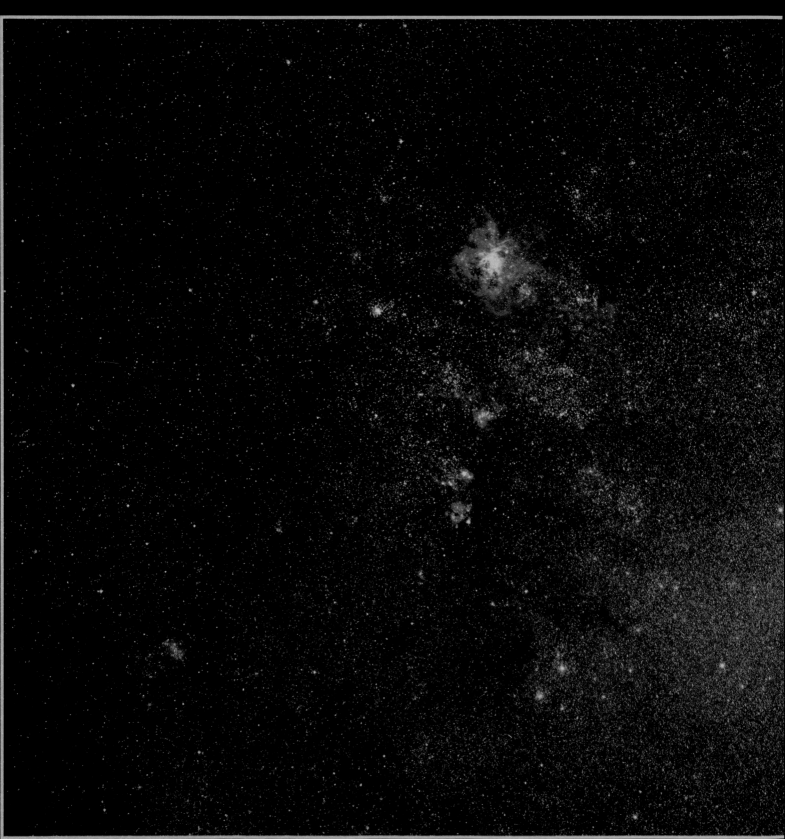

Nearest galaxy to Earth, the Large Magellanic Cloud glows pink from billows of energized hydrogen, including the spidery Tarantula nebula *(top left)*, a vast accumulation of gas that spans more than 6,000 light-years.

lue and lovely in its gauze of clouds, the Earth spins its way through a universe huge beyond imagining. Every night, the star-dusted cosmos reveals itself overhead, and every night since the dawn of the human species, people have looked up and wondered about its extent and workings. Ancient astronomers saw gods and mythic creatures patterned in the heavens and surmised that humankind's home stood at the center of creation. But eventually the humbling truth became clear. The Earth turned out to be one of nine planets circling the Sun, and the Sun itself proved to be just one of more than one hundred billion stars in a galaxy—a stellar multitude bound together by gravity.

Time and gravitational effects have shaped that multitude into a great, glowing disk. In the night sky, the star-thronged plane of the disk forms a pale river of light that gives the galaxy its name: the Milky Way. (*Galaxy* comes from the Greek for "milk.") So vast is the Milky Way that if it were reduced to the size of the United States, the Earth would be far smaller than the smallest dust mote, barely visible through an electron microscope.

Only a few decades ago, astronomers gained proof of what some of them had long suspected—that the Milky Way does not begin to embrace all that there is. The cosmos stretches far beyond the edges of the Sun's galactic home, to regions barely guessed at. Charting that immensity has been one of the great scientific endeavors of the twentieth century. Night after night, small cadres of men and women bundle themselves against the cold and slide open the doors of their domed observatories. From these mountain shelters emerge a host of instruments used to investigate the depths of the universe. The Milky Way, astronomers have discovered, is only one among uncounted hundreds of billions of other galaxies, each a prodigious star system in its own right. The diameters of the largest are three times that of the Milky Way. Even so-called dwarf galaxies, with diameters about one-thirtieth our galaxy's span, are capacious enough to hold a hundred thousand stars.

In their placement and behavior as well as their size, they are a disparate lot. Some galaxies pass their lives in relative isolation, hanging jewel-like in the void. Others occur in large aggregations: Studies of certain parts of the sky reveal tens of thousands of them in an area that could be covered by a hand held out at arm's length. (Scaled down to the size of people, galaxies typically are about as widely separated as players on a rather extended baseball field.

In contrast, if the stars that make up a galaxy were people size, the nearest neighbors would live 60,000 miles away.)

Some galaxies are serene; others spew out energy or matter with inconceivable violence. The strangest among them, lying at the fringes of the observable universe, are known as quasars, which seem to be the size of mere solar systems but generate a hundred times the energy of a typical galaxy. Even the Milky Way, one of the quieter systems, evidently harbors a powerful energy source at its center, where the outpouring of radiation may be fueled by one of the most intriguing of cosmic phenomena: a black hole.

In structure, too, galaxies display great variety. Perhaps three-quarters are disk shaped like the Milky Way; because their stars seem wound about their centers, they are termed spiral galaxies. Others, called ellipticals, are more spherical. And still others have been wrenched into asymmetry by the gravitational tug of galactic neighbors.

THE GALACTIC NEIGHBORHOOD

In all the heavens, only three galaxies may be easily perceived with the naked eye. The trio belongs to the so-called Local Group *(pages 16-17)*, a cluster of thirty or more galaxies that includes the Milky Way. Two of them are known as the Large and Small Magellanic Clouds, named for the sixteenth-century Portuguese explorer Ferdinand Magellan. Navigators aboard his ships saw them as a pair of luminous, cloudy patches in the skies of the Southern Hemisphere; modern photographs reveal their embrace of billions of stars. Their light, fleeting across space at the rate of 186,000 miles per second, takes 170,000 years to reach the Earth from the larger cloud, and 190,000 years from the smaller. The third member of the trio lies in northern skies in the constellation Andromeda. Because its light takes more than two million years to reach the Earth, we see this galaxy as it was when the first protohumans were learning to walk upright.

These three are among our closest galactic neighbors—virtually next door by astronomical standards. Yet, along with hundreds of millions of other such star systems discernible with telescopes, they remained cryptic smudges of light until just a few decades ago. Only in the mid-1920s did most astronomers fully accept the existence of galaxies beyond our own star swarm. The reason that recognition took so long could be summed up in a single word: distance. To generations of sky watchers, the pinpricks of radiance in the night sky were a kind of two-dimensional tale, an image utterly lacking in depth. Until astronomers developed ways of determining what was far and what was near, what was big and what was small, the heavens held much more mystery than they did fact.

Ironically, the first tentative step toward the discovery of galaxies and the scale of the universe was taken by a man whose interests did not extend much beyond the Solar System. The eighteenth-century French astronomer Charles Messier was a comet hunter, ever on the lookout for those fascinating wanderers in the solar family. During his lifetime, Messier claimed to have dis-

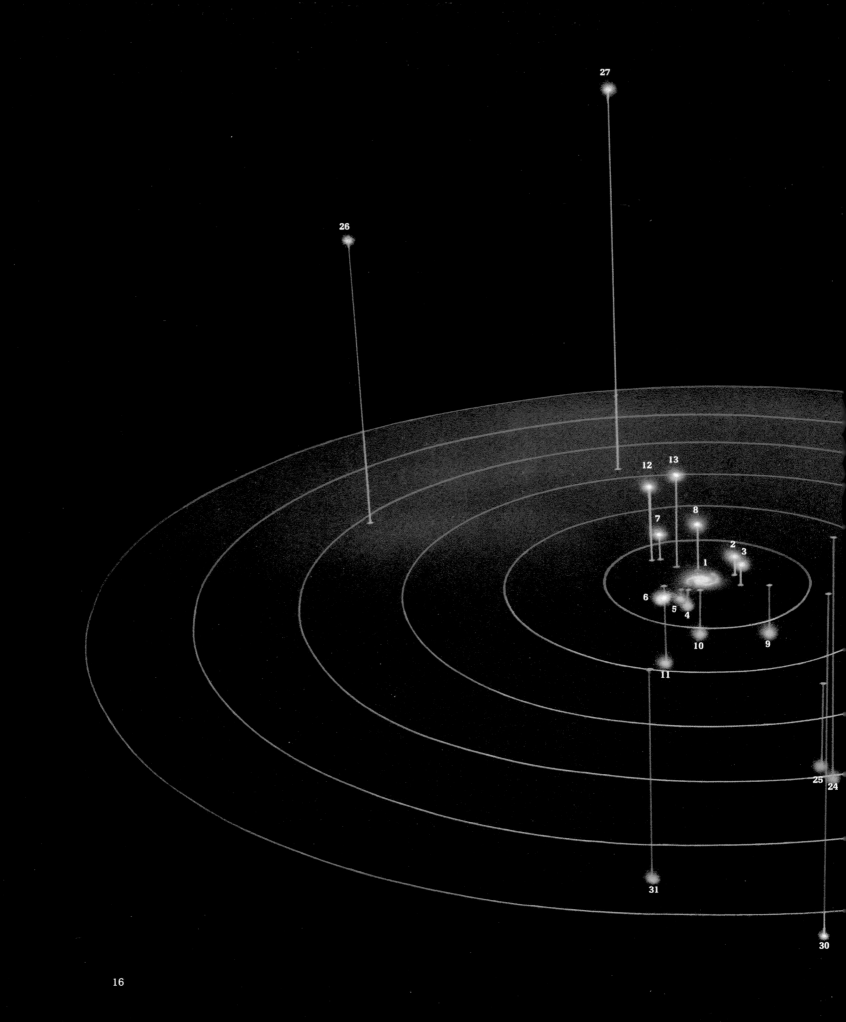

The numbering of galaxies on this map of the Local Group begins in the center with the Milky Way and moves outward clockwise. The circular blue lines indicate intervals of roughly 650,000 light-years.

(1) Milky Way. (2) Draco. (3) Ursa Minor. (4) Small Magellanic Cloud. (5) Large Magellanic Cloud. (6) Carina. (7) Sextans C. (8) Ursa Major. (9) Pegasus. (10) Sculptor. (11) Fornax. (12) Leo I. (13) Leo II. (14) Maffei I. (15) NGC 185. (16) NGC 147. (17) NGC 205. (18) M32. (19) Andromeda I. (20) Andromeda III. (21) Andromeda (M31). (22) M33. (23) LGS 3. (24) IC 1613. (25) NGC 6822. (26) Sextans A. (27) Leo A. (28) IC 10. (29) DDO 210. (30) Wolf-Lundmark-Melotte. (31) IC 5152.

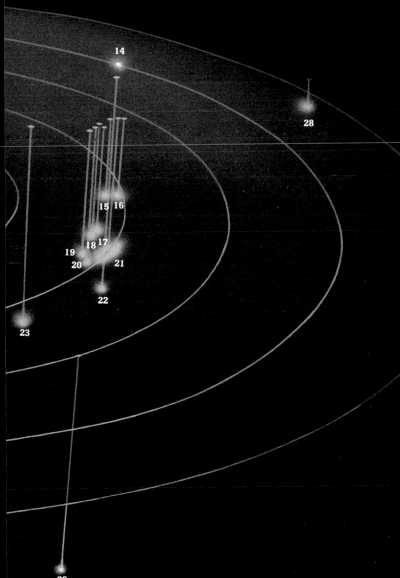

The precise membership of the group is uncertain, for several reasons. Small galaxies at even a moderate distance can blend into the background celestial glow, and quite large galaxies could be hidden by the Milky Way itself. The roster is also subject to dispute because astronomers differ on whether faint galaxies whose distances are in question should be considered part of the Local Group.

In the map at left, the generally accepted members are located around an imaginary plane centered on the Milky Way (1). The galaxies tend to have cryptic-sounding names. Most are simply the galaxies' first listing in an astronomical catalog: "M" numbers, for example, refer to a catalog assembled by the eighteenth-century French astronomer Charles Messier; "NGC" numbers refer to the *New General Catalogue* compiled in the nineteenth century. So-called dwarf galaxies, which are smaller systems of more recent discovery, have often been named after the constellation in which they appear—Leo I and Leo II (12, 13), for instance. A few, such as Wolf-Lundmark-Melotte (30), commemorate the scientists who found them.

The Milky Way and the Andromeda galaxies (21), giant spirals whose disks span more than a hundred thousand light-years, dominate the group's irregular formation. Separated by more than two million light-years, each serves as the focus for the orbits of several considerably smaller satellites. Andromeda's nearest companions, for instance, are each only about a hundredth the mass of the larger galaxy. A handful of more-isolated galaxies reside on the fringes of the group. Some—like M33 (22)—are relatively large, perhaps a tenth the mass of the Milky Way. But most are as diminutive as the satellite star systems.

covered twenty-one comets, although by modern standards of discovery probably only fifteen or sixteen can in fact be attributed to him. Whatever the number, King Louis XV referred to him as the "comet ferret."

Messier's methods were much the same as those employed by amateur astronomers today, who still find many of the ten or so new comets discovered each year. He would systematically scan the heavens with his telescope (the largest he ever worked with had an effective diameter of less than eight inches) until he spotted a faint blurry patch, the sign of a comet just beginning to glow in the light of the Sun. He would then check its position against a star chart and continue the search. The next night he would look to see if the patch had moved. If it had, the object was probably a comet; over time its orbit could be calculated so that he could determine if it had already been discovered. If the glimmer had not budged, it was something else—not a comet, not a star, but simply an inexplicable fuzzy spot, fixed in the heavens.

Already by Messier's era these spots were known as nebulae, from the Latin word for "mist" or "clouds," and Messier found them a great annoyance— false trails that disrupted the hunt. So he began to keep track of them, and in 1774, for the convenience of other comet hunters, he published a list of forty-five nebulae and their celestial coordinates. Over the next decade he issued supplements to his list and in 1784 published a catalog of 103 objects. Today Messier is remembered not for his cherished comets but for the nebulae he so assiduously sought to avoid.

A MUSICIAN-ASTRONOMER

Not all of Messier's contemporaries shared his lack of interest in the nature of the nebulae or in broader cosmic questions. The most prominent sky watcher to examine the nebulae more closely was William Herschel, a German-born musician who emigrated to England as a teenager in 1757. Herschel did not take up astronomy until he was thirty-five years old, but in the second half of his life, he became the world's most renowned stargazer, discoverer of the planet Uranus and court astronomer to King George III.

In some ways Herschel was a man ahead of his time. He was the first to build telescopes comparable in size to those used by astronomers today: His largest reached forty feet in length and had a mirror forty-eight inches in diameter. (However, it was so unwieldy that Herschel more often used an instrument half its length.) He was also the first bold enough to make a systematic attempt to ascertain the size and structure of the universe.

Aided by his sister, Caroline, who would become an accomplished astronomer in her own right, Herschel undertook a comprehensive survey of the sky, setting out to observe every item on Messier's list. Because he was using a telescope with more than four times the light-gathering power of Messier's largest instrument, he fully expected to find nebulae that Messier had missed. Over the next seven years, he discovered 2,000.

The methodical Herschel's next objective was to construct a map of these bountiful heavens. From his study of the nebulae, he had formed the opinion that, if viewed from a great enough distance, the Milky Way itself would resemble a nebula. Furthermore, he believed he could determine the Milky Way's dimensions by estimating in each direction the numbers of stars of various brightnesses. The key to the effort was an assumption that all stars are intrinsically as bright as the Sun; dimness therefore would be a measure of their distance. In this he was wrong, as he himself eventually realized. He gave up trying to take the measure of the galaxy but held to his conviction that the Milky Way would look like a nebula to an observer very far away.

OBSERVATIONS FOR AND AGAINST

Over the course of his career, Herschel's speculations about the nature of nebulae also underwent changes. His telescopes revealed that many of Messier's fuzzy objects could be resolved into clusters of stars, some of which came to be known as globular clusters for their shape. He took this as evidence that supported the so-called island universe theory. The term had been coined two decades earlier by the philosopher Immanuel Kant in a remarkably intuitive commentary on astronomical writings he had come across. According to the theory, any nebulae that could not be resolved into aggregations of stars, even through the most powerful telescopes of the day, must be separate island universes—galaxies at such great distances that individual stars cannot be seen. Herschel, having been able to resolve some nebulae into stars, was inclined to accept the idea that those he could not resolve must indeed be island universes. But then he found a nebula made up of a single star surrounded by "a faint luminous atmosphere, of a circular form." (It was, in fact, a so-called planetary nebula, a shell of gas around a dim star that seemed to some astronomers to bear a superficial resemblance to a planet.) This atmosphere was so clearly linked with the star that either the star had to be a giant of enormous proportions or the nebulosity was made up of extremely tiny stars. Rejecting both options, he concluded that the central body was an ordinary-size star "involved in a shining fluid, of a nature totally unknown to us." Although he was delighted by the possibilities this novelty opened up, the mystery of the nebulae was merely compounded.

Herschel was the only astronomer of his day to work with telescopes large enough to observe the nebulae in any detail. The next astronomer to tackle the riddle was an Irish nobleman, William Parsons, the third earl of Rosse. In the early 1840s, Parsons began building a telescope that would outdo even Herschel's masterpiece. Fifty-four feet long, with a metal mirror six feet in diameter, it was known as the Leviathan of Parsonstown, after the location of Parsons's country seat in Ireland. The telescope's enormous mirror would not be surpassed for three-quarters of a century.

Toward the end of the decade, Parsons turned the Leviathan toward the object known as M51 (the designation in Messier's catalog of nebulae). He beheld a remarkable, almost bewitching sight. The nebula displayed a spiral pattern, a shape that later earned it the name Whirlpool. Parsons soon began to find additional spiral-shaped nebulae in the sky, and he made every effort to pick out stars in them. But despite the great light-gathering power of the Leviathan, no stars could be seen. The spiral nebulae could only be classed as a lovely conundrum.

THE FIRST BREAKTHROUGH

Finding stars in the spirals would have been a significant event but would not, on its own, have answered a more pressing question: Were these nebulae part of the Milky Way? The answer depended on solving two related puzzles, the size of the Milky Way itself and the distance to the nebulae. Neither solution would be possible until astronomers could find a reliable celestial yardstick.

Partial clues began to appear. A few years before Parsons began building the Leviathan, astronomers finally succeeded in determining the distance to a star. They used a method called trigonometric parallax, which bears many similarities to the way the eyes determine relative distance. Because a person's eyes are an inch or so apart, each eye views the world from a slightly different perspective. Looked at with first one eye and then the other, an object in the foreground—a pencil held at arm's length, say—seems to shift a bit against a stationary background, such as a wall. In astronomy, two positions of Earth on opposite sides of its orbit around the Sun correspond to the eyes; very remote stars make up the unchanging backdrop against which the parallax (and hence the distance) of considerably nearer stars may be measured *(left)*. So far away are even nearby stars, however, that their parallaxes are vanishingly small—less than the apparent width of a nickel at three miles. Beyond about 600 light-years, measurement becomes grossly inaccurate. As for the nebulae, none showed any evidence of parallactic motion; their distances remained a cosmic secret.

But the measurement of stellar parallaxes, though inadequate to solve the mystery of the nebulae directly, did give astronomers a new tool for learning more about nearby stars, and from this base of knowledge they would begin to reach deeper into the universe.

CLUES IN THE LIGHT OF THE SUN

Another technique developed in the nineteenth century—spectroscopy—proved to be an even more potent springboard. By the early 1800s, scientists knew that sunlight flares into a multihued spectrum when passed through a glass prism. They also had learned that if the sunlight is directed through a narrow slit before entering the prism, the resulting spectrum will be interrupted by dark lines and bands. Several decades later, it became clear that these lines were caused by the absorption of particular wavelengths as light passed through various gases in the Sun's atmosphere. Laboratory experiments revealed which gases absorbed which wavelengths. (A hot gas emits certain wavelengths of light in the form of bright emission lines; it absorbs those same wavelengths from light passing through it.) It was now possible to begin answering a question that would have seemed preposterous before spectroscopy: What is the Sun made of?

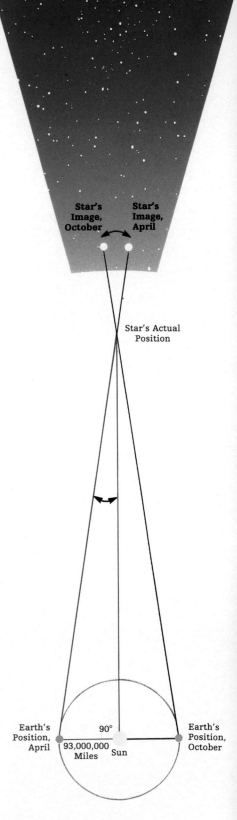

Star's Image, October

Star's Image, April

Star's Actual Position

Earth's Position, April

90°

93,000,000 Miles

Sun

Earth's Position, October

20

But the full utility of spectroscopy had barely been tapped. One of the first people who thought to turn a spectroscope-equipped telescope on a star other than the Sun was William Huggins, a gentleman astronomer who had built an observatory on top of his London home. In 1868 Huggins performed studies that applied a theory put forth some twenty years earlier by Christian Doppler, an Austrian mathematician and physicist. The theory was later significantly refined by physicist Armand Fizeau of France.

The so-called Doppler effect (Fizeau has, for the most part, been unfairly neglected) causes all the wavelengths of light emitted by an object to be shifted toward the shorter-wavelength end of the visible spectrum (blue) when the object is approaching Earth and toward the longer-wavelength end (red) when it is receding. The object itself does not necessarily change color, but individual wavelengths in its spectrum, such as those represented by the absorption lines, should change position relative to the spectrum of a similar object at rest *(pages 22-23)*.

Huggins searched for such a shift in the spectrum of the brightest star in the sky—Sirius, the Dog Star. He found one: a slight shift to the red. From this he calculated that Sirius has a velocity of thirty miles per second away from the Earth along the line of sight. Huggins applied this technique to several other stars and found similar speeds, with some stars receding and some approaching. By the end of the century, the measurement of line-of-sight velocities would occupy astronomers at several observatories. But two more decades would pass before the Doppler effect would assume its true value in astronomy—as a versatile device for judging distances vastly greater than those measurable by parallax.

AMBITIOUS ANALYSIS

In the meantime, first spectroscopy and then photography came into their own as part of the standard astronomical tool kit. As the twentieth century began, the Harvard College Observatory, under director Edward C. Pickering, was working on several ambitious star surveys. To perform the tedious job of taking measurements and making calculations from the observatory's thousands of photographic and spectroscopic plates, Pickering hired local women, who were willing to work for less money than men and were thought to be temperamentally better suited to the job. The women, dubbed "Pickering's harem" by later astronomers, generally went unrecognized by the astronomical community, yet several would make major contributions to the field. One of the most gifted was Henrietta Swan Leavitt, who would lay the groundwork for a revolutionary series of distance-measuring methods.

Leavitt, the daughter of a Congregational minister, was afflicted with deafness and reserved in manner. She was also unmistakably brilliant. While attending what would become Radcliffe College, she became interested in astronomy, and after graduation in 1892, she joined the Harvard College Observatory as a volunteer research assistant. In 1902 she took a permanent position, soon rising to become head of a department that specialized in

A Stellar Distance Gauge. The astronomical term *parallax*—from the Greek word for "change"—refers to the way in which the position of a nearby star seems to shift against the motionless background of very remote stars as Earth orbits the Sun. By simple trigonometry, that shift can be used to calculate the star's distance.

In the example at left, the star is photographed during the month of April and then six months later, when Earth has traveled halfway around the Sun. The star's apparent movement across the celestial background is measured on the photographs as an angle, greatly exaggerated here. Halving that measurement yields one angle of a right triangle *(red)* that has the star, the Sun, and Earth at its vertices. Because the Sun-to-Earth distance is known, the leg of the triangle representing the star-to-Earth distance is easily determined. With ground-based instruments, the technique is reliable only out to about 600 light-years.

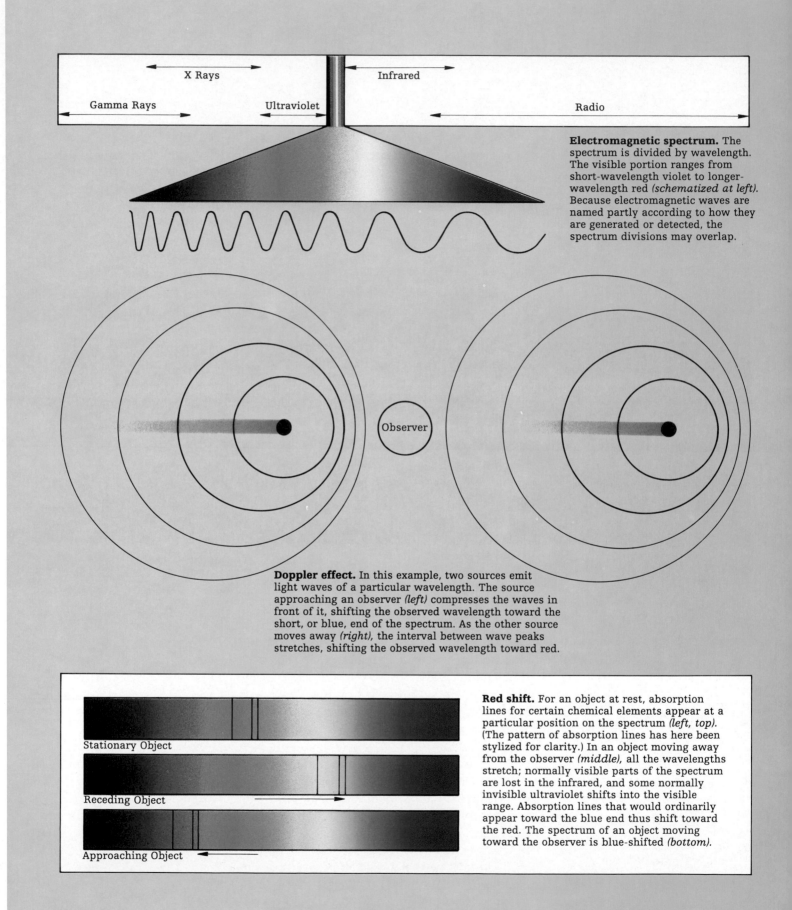

Electromagnetic spectrum. The spectrum is divided by wavelength. The visible portion ranges from short-wavelength violet to longer-wavelength red *(schematized at left)*. Because electromagnetic waves are named partly according to how they are generated or detected, the spectrum divisions may overlap.

X Rays

Infrared

Gamma Rays

Ultraviolet

Radio

Observer

Doppler effect. In this example, two sources emit light waves of a particular wavelength. The source approaching an observer *(left)* compresses the waves in front of it, shifting the observed wavelength toward the short, or blue, end of the spectrum. As the other source moves away *(right)*, the interval between wave peaks stretches, shifting the observed wavelength toward red.

Stationary Object

Receding Object

Approaching Object

Red shift. For an object at rest, absorption lines for certain chemical elements appear at a particular position on the spectrum *(left, top)*. (The pattern of absorption lines has here been stylized for clarity.) In an object moving away from the observer *(middle)*, all the wavelengths stretch; normally visible parts of the spectrum are lost in the infrared, and some normally invisible ultraviolet shifts into the visible range. Absorption lines that would ordinarily appear toward the blue end thus shift toward the red. The spectrum of an object moving toward the observer is blue-shifted *(bottom)*.

Red Shift: A Spectral Speedometer

Celestial bodies radiate energy at a wide range of wavelengths along the electromagnetic spectrum, including the slender segment comprising visible light *(left, top)*. For astronomers, light has many tales to tell, not least those writ by the so-called Doppler effect *(middle)*. This simple rule of physics states that the observed frequency of a light wave is affected by the relative motion of source and observer, just as the pitch of a siren changes from high to low as it passes.

Stellar spectra vary with chemical composition. As light generated within a star passes through the star's gaseous outer layers, certain wavelengths are absorbed by various gases, producing a distinctive pattern of dark absorption lines in the rainbow-hued spectrum *(bottom)*. A galaxy's spectrum shows these lines as well, since it is a composite of the spectra of its billions of stars. Because of the Doppler effect, the motion of a celestial object will cause absorption lines to shift position along the spectrum. These shifts are very small, often representing a change in wavelength of only a few millionths of a millimeter. But by comparing the positions to an unshifted set of lines produced in a laboratory, astronomers can learn two critical facts: whether the object is moving toward or away from the Earth and how fast it is traveling.

measuring the brightness—or magnitude—of stars as recorded on photographic plates. (The scale of magnitudes, explained on page 24, is designed to let astronomers easily compare stars of differing brightness.)

Her work centered on a set of plates produced over a period of years by a twenty-four-inch telescope that Harvard operated in the mountains of Peru. The plates depicted a swarm of stars familiar to observers in the Southern Hemisphere—the nebula known as the Small Magellanic Cloud. In the cloud, Leavitt noticed numerous examples of a pulsating type of star called a Cepheid variable, so named because the first such cosmic firefly to be identified was the star Delta Cephei, in the constellation Cepheus. For reasons still not fully understood, Cepheid variables expand and contract with regularity; as a result, they blaze brightly, then dim, then blaze up again as the cycle repeats. A Cepheid variable's period, or cycle time, may be as short as a day or as long as many months. Whatever the rhythm, it generally persists with metronomic constancy—accurate to within fractions of a second.

A SIGNIFICANT RELATIONSHIP

By 1908 Leavitt had compiled a list of well over a thousand such variables in the Small Magellanic Cloud. Sixteen of them appeared on enough different photographic plates to enable her to determine their periods. After studying them in detail, she noticed a curious feature: the longer the periods, the brighter the stars. In 1912, having widened her examination to a total of twenty-five Cepheids, she published a paper showing that the period of Cepheids and their observed brightness—"photographic luminosity," as she called it—were mathematically connected: A particular period, or rate of pulsation, was always associated with a particular maximum brightness on the photographic plates. All the Cepheids could be fitted onto a single scale, a graph defining a smooth relationship between period and luminosity.

Discovery of the relationship was an epochal moment in astronomy. It offered a way around the frustrating fact that—as William Herschel had realized a century before—a star's observed brightness, or apparent magnitude, cannot by itself indicate distance because it is a product of both the actual output of light and how far that light travels before reaching Earth. (The dimming effects of distance are governed by the so-called inverse-square law: A star, like any light source, will look a quarter as bright if its distance from the observer is doubled, a sixteenth as bright if the distance is quadrupled, a sixty-fourth as bright if the distance is multiplied by eight.)

Leavitt had linked the apparent magnitude of variable stars to one measurement that is not influenced by, or dependent on, the stars' distance—namely, the rate at which their brightness changes. She had then made a vital connection between the stars' periods and their yet-unknown absolute magnitudes—their actual luminosities. "Since the variables are probably at nearly the same distance from the Earth," she wrote, "their periods are apparently associated with their actual emission of light."

This was so because all the stars in the Small Magellanic Cloud clearly

A System for Measuring Brightness

Among the most fundamental pieces of information about stars and other celestial bodies, particularly for estimating cosmic distances, are measures of brightness. Because brightness decreases with the square of the distance—the so-called inverse-square law—astronomers have devised a convention that allows them to compare objects of different intrinsic luminosities, regardless of how far away they are. The measures they use are known as *apparent magnitude* and *absolute magnitude,* respectively.

The magnitude scale dates from work done in the second century BC by the Greek philosopher Hipparchus. He labeled the brightest stars he could see "first magnitude," the next brightest "second magnitude," and so on down to sixth magnitude, which was the faintest star that could be seen with the unaided eye. Thus, the lower a star's magnitude number, the brighter it is.

Later astronomers determined that the difference in brightness between one magnitude and the next was always an equal ratio. A star of magnitude 1 is about 2.5 times brighter than a star of magnitude 2, which in turn is 2.5 times brighter than one of magnitude 3; a first-magnitude star is thus about 100 times brighter than a star of magnitude 6. These scientists also found that some of the stars Hipparchus deemed first magnitude were actually brighter than that, and decreed that stars could have zero and negative magnitudes. For example, the brightest star in the sky, Sirius, has a magnitude of -1.45. On the other side of the coin, modern telescopes can detect objects of the twenty-fourth magnitude, nearly 16 million times fainter than Hipparchus's sixth-magnitude stars.

All of these measures refer to the objects' *apparent* magnitudes—their observed brightnesses as recorded on Earth. To compare objects of different intrinsic luminosities, astronomers must somehow compensate for the effects of distance. What they do, in essence, is pretend to place stars and galaxies at an arbitrary standard distance (32.6 light-years), rather like lining up light bulbs of different wattages. To do this, they need two things: an object's apparent magnitude, taken from photographic plates, and a reasonable estimate of its actual distance. Then they take the distance a given object would have to be moved in—or pushed out—to reach 32.6 light-years and plug that distance, as well as the object's apparent magnitude, into the inverse-square relation.

The result is a measure of how bright the object would appear if viewed at the standard distance. By convention, this is deemed its *absolute* magnitude. Obviously, when very dim objects hundreds of light-years away are brought to within 32.6 light-years of Earth, their absolute magnitudes will be considerably brighter than their apparent magnitudes. The reverse would be true of very bright—but nearby—objects. The Sun, for example, is eight light-minutes away and has an apparent magnitude of -26.7. If it were pushed out to 32.6 light-years, it would be as dim as a fifth-magnitude star.

belonged to the same remote celestial structure. They were therefore equally distant from Earth in the same sense that everyone in Boston can be considered equally distant from London. Thus, if one of the Magellanic Cepheids appeared four times as bright as another, it really was four times as bright—and not, for example, half as far away.

A SIMPLE MATCH TRICK

Her chain of logic led in a fascinating direction. If the absolute magnitude of a nearby Cepheid could somehow be measured, that same magnitude could be ascribed to a Cepheid with a matching period in the Small Magellanic Cloud. Then—and here was the payoff—the difference between the Magellanic Cepheid's absolute magnitude and its apparent magnitude could be used to calculate the star's distance simply by allowing for the dimming effects of the inverse-square law. And that was just the beginning. Because Leavitt had discovered a mathematical relationship that applied to the Magellanic Cepheids in general, finding the absolute magnitude of one would allow astronomers to infer the absolute magnitudes of all of them. The period-luminosity scale could then be used to find the absolute magnitude of any similar variable, wherever it occurred, by matching its period to the scale. The distance to that variable, and to any celestial body associated with it, was then a straightforward calculation.

Henrietta Leavitt was unable to pursue the promise of her period-luminosity formula, however. Edward Pickering, her boss at the observatory, assigned her to other tasks involving photographic measurement techniques. He believed that the observatory's job was to collect data, not to chase after its cosmic meanings. In any event, a serious obstacle stood in the way of creating a celestial distance-gauge out of Cepheid behavior: The nearest Cepheid to Earth lies well beyond the reach of parallax measurement.

But a young Danish astronomer named Ejnar Hertzsprung, then working in an observatory near Berlin, read Leavitt's paper and concluded that the needed distance-figure could be derived by a different technique, one that focused on a type of stellar movement known as proper motion. The proper motion of a star is a complex phenomenon made up of the movements through space of both the star and our Sun. Measurement of such movements requires analysis of photographic plates taken years or decades apart and is only approximate for any individual star. But taking a statistical sample of the proper motions of several stars can yield an idea of the stars' distance.

Hertzsprung found this type of information for thirteen Cepheid variables in the Sun's neighborhood. By statistically combining the data, he worked out, in effect, an "average" distance for the local Cepheids. He also computed an average apparent magnitude. Using these values and applying the inverse-square law that took the dimming effects of distance into account, he then calculated an average absolute magnitude for a Cepheid of an average period.

The next step was to find a Cepheid in the Small Magellanic Cloud with the same period—and presumably the same absolute magnitude—as his mathematical "average" star. Lastly, he compared the photographic brightness of the Magellanic Cepheid to its presumed absolute magnitude and computed the distance to the cloud. It was a clever piece of work by a great astronomer. (Hertzsprung would later make fundamental contributions to the understanding of stars.) But Hertzsprung's results, published in 1913, were not very exciting: He estmated the distance of the Small Magellanic Cloud at a mere 3,000 light-years, well within the presumed boundaries of the Milky Way. That figure may, in fact, have been a mistake; a zero evidently had been dropped, perhaps through a typographical error. Hertzsprung's methods should have led him to a distance estimate of 30,000 light-years.

Even at 30,000 light-years, however, Hertzsprung's calculation would have been off the mark by as much as 160,000 light-years, for a complex of reasons. In part the large discrepancy was due to the fact that the Cepheids in the Small

Magellanic Cloud had been photographed with plates sensitive to blue light, whereas the local Cepheids had been photographed with red-sensitive plates, and there was no way to make an exact correction for the difference in apparent brightness. Hertzsprung's guess had the effect of making the Magellanic stars seem brighter—and thus closer—than they actually are.

This difficulty was unappreciated at the time, but most astronomers saw other flaws. They felt that thirteen Cepheids was a trifling number to use as the basis of the distance estimate. Perhaps more important, not everyone was convinced that Leavitt's relationship for variables in the Small Magellanic Cloud applied to Cepheid variables in the Milky Way. But at least one other individual recognized the value of Cepheids, and with them he would radically change the prevailing view of the universe.

FROM MISSOURI TO MOUNT WILSON

Harlow Shapley might have seemed an unlikely person to become one of the most prominent astronomers of the twentieth century. Raised on a Missouri farm, he had a sketchy early education and, while still a teenager, worked as a crime reporter for small-town newspapers. But he picked up a few high-school courses and was accepted at the University of Missouri, where he planned to enter Missouri's new school of journalism. This scheme came unraveled when the opening of the journalism school was put off for a year. He cast about for another field of study and settled on astronomy.

A quick and self-confident student, Shapley proved adept at his chosen field, winning a fellowship to Princeton. There he studied variable stars, including Cepheids. The caliber of his work

1848 William Parsons, third earl of Rosse, confirmed a spiral shape in M51, now called the Whirlpool galaxy.

1802 William Herschel culminated years of observation with catalogs expanding the list of nebulae to 2,500.

1784 Charles Messier published a catalog of 103 nebulae—including 33 later found to be galaxies.

A LONG ASTRONOMICAL PURSUIT

One of astronomy's longest-running controversies centered on the cloudlike nebulae seen among the stars. Some astronomers considered them part of the Milky Way galaxy, itself thought to be the whole universe. Others, believing in a smaller Milky Way, viewed the nebulae as galaxies in their own right. The issue hinged on finding accurate distances and was not resolved until the 1920s. Depicted here are some of the key figures whose work provided clues that helped determine the Milky Way's place in the cosmos.

1842 Christian Doppler *(upper left)* discovered the effect of motion on sound waves; in 1848, Armand Fizeau *(left)* showed that the Doppler effect would cause the lines in a star's spectrum to shift.

brought him the offer of a job at Mount Wilson Observatory, overlooking Los Angeles. At the time, Mount Wilson was on its way to becoming the most sophisticated observatory in the world. When Shapley arrived in 1914, he had at his disposal a recently completed sixty-inch telescope. A few years later, the observatory's director, George Ellery Hale, installed a 100-inch telescope—the largest in existence for thirty years.

The power of the sixty-inch was intoxicating. Shapley began studying the glittering stellar assemblages known as globular clusters, which Messier had classified as nebulae and which Herschel and Parsons had found to be spherical gatherings of stars. When Shapley turned the sixty-inch telescope on the globular clusters, he discovered Cepheid-type variables. These Cepheids had shorter periods than the ones used by Hertzsprung, but Shapley did not see that as a problem. Using Hertzsprung's calibration—and later refining it—he determined the distances to the nearest clusters, then used those figures to develop additional distance indicators, known as standard candles. For instance, he systematically compared the thirty brightest nonvariable stars in a cluster (discarding the first five on the assumption that they might be stars in the foreground) with Cepheids in the same cluster. The Cepheids' periods revealed their absolute magnitudes, which in turn gave him an idea of the absolute magnitudes of the brightest nonvariable stars. He then made the assumption that the brightest stars in all clusters are very much alike, allowing him to use them as standard candles to find the distance to clusters with no visible Cepheids. In this way, he calculated the distances of all the clusters he could find.

The numbers he came up with were astonishing. At the time, some astronomers thought the Milky Way was about 30,000 light-years across, but there was no consensus, and many opted for a number as small as a few thousand. According to Shapley, the clusters were 50,000 to 220,000 light-years away.

If they were part of the Milky Way—and there was no reason to assume otherwise—the Milky Way was perhaps ten times larger than anyone had suspected. Using his distances to the globular clusters, Shapley calculated a

1868 William Huggins found a noticeable red shift in the spectrum of Sirius and calculated the speed of the star's recession from the Earth.

1912 Henrietta Leavitt, studying Cepheid variable stars in the Small Magellanic Cloud, noted a direct relation between their luminosity and their period of fluctuation.

1913 Ejnar Hertzsprung derived an average distance from Earth for Cepheids in the Milky Way galaxy, then used Leavitt's period-luminosity scale to calculate—incorrectly, as it turned out—the distance to the Small Magellanic Cloud.

new diameter for the galaxy of 300,000 light-years, with the galactic center located in the direction of the constellation Sagittarius, where more than a third of the globular clusters he studied were congregated. He published what he called his Big Galaxy theory in 1918, but questions and protests greeted him from every quarter. For the most part, astronomers were as wary of Cepheids then as when Hertzsprung tried them out in 1913.

Controversial as they were, Shapley's estimates had important implications for the ongoing debate over the nature of the spiral nebulae. Despite the continuing inability of telescopes to discern stars in spiral nebulae, many astronomers in the early years of the twentieth century suggested that they were galaxies comparable to the Milky Way. For one thing, their light, when passed through a spectroscope, resembled that of stars, not clouds of gas.

Furthermore, by 1917 astronomers comparing photographs of spirals had begun to find telltale objects called novae (from the Latin for "new") in several of them. Novae are stars that undergo recurrent eruptions of matter, radically changing in luminosity. (A supernova, in contrast, is a one-time explosion of an entire massive star.) A single Milky Way nova might accidentally appear directly in line with a spiral, but for several to do so was highly improbable.

The implications were significant: The possibility that the novae were actually part of the spiral nebulae constituted the first hint that these long-unresolvable objects might be stellar congregations. Moreover, if the novae in the spirals were the same type of star as those found in the Milky Way, the fact that they were, on average, ten magnitudes fainter than Milky Way novae meant that they were at least a hundred times farther away.

Some astronomers began using novae as indicators of distance to the spirals and found values that placed the nebulae well beyond the Milky Way (even at its largest accepted size of 30,000 light-years). Not everyone was persuaded by the argument, of course; perhaps the lights represented collisions of gaseous nebulae with small dark stars, or perhaps they were some entirely new phenomenon.

Early in his career, Shapley favored the notion of island universes, and when novae were found in spirals, he was in the forefront of those who supported them as evidence that spirals were separate galaxies. However, in formulating his Big Galaxy theory, he changed his mind. The Milky Way as he envisioned it was much too large for all of the

1914 Vesto Slipher announced Doppler-shift measurements for fifteen spiral nebulae; thirteen indicated vast speeds of recession. However, these large red shifts remained an enigma for many years.

1920 In the Great Debate, Harlow Shapley *(left)* used Cepheids as distance indicators to fix the Milky Way's diameter at 300,000 light-years; he argued that the spiral nebulae were small systems associated with the galaxy. Heber Curtis *(below)* believed the Milky Way to be one-tenth that size, the nebulae to be galaxies themselves.

spirals to be comparable. For that to be the case, they would have to be even farther away than the novae indicated. Shapley's new conviction led to a confrontation that would become legendary—although it passed virtually unremarked at the time.

In 1920 the National Academy of Sciences sponsored a debate called "The Scale of the Universe" at its annual meeting in Washington, D.C. The debaters were Shapley and Heber D. Curtis, who worked at the Lick Observatory, southeast of San Francisco. The astronomers at Lick were opposed to the Big Galaxy theory—to such an extent that Shapley was known to refer to their skepticism as the "Lick state of mind." Instead, they championed the island universe theory. Curtis was an articulate and powerful public speaker. He was also the first person to have found novae in the arms of spiral nebulae, although his characteristic caution in delaying an announcement allowed another astronomer, George Ritchey, to get credit for being the first.

The oral portion of what has come to be known as the Great Debate was a rather lackluster affair. In fact, the two presentations seemed to deal with separate topics. The debate continued—some would say the real debate took place—about a year later, in the pages of the *Bulletin of the National Research Council,* where both men took pains to spell out their positions.

TAKING AIM AT THE VARIABLES

In both his oral and written arguments, Curtis concentrated on what he saw as the weakest link in Shapley's theory—the Cepheid variables. Shapley had based his view of the universe on the same thirteen Milky Way Cepheids that Hertzsprung had used; in fact, he had thrown out two of the thirteen because he mistrusted the data from them. This might not have been telling in itself, but Shapley had also disregarded the fact that the Cepheids he found in the globular clusters had shorter periods than their local counterparts. Although Shapley extended Leavitt's period-luminosity relationship to accommodate these shorter-period Cepheids, Curtis argued that it might not apply to them.

Curtis then turned to the question of the spiral nebulae, mentioning his own work on novae. He also subscribed to another finding that he felt clinched the status of the spirals as separate galaxies. In 1912, at the Lowell Observatory in Flagstaff, Arizona, a skillful and patient spectroscopist named Vesto Slipher had succeeded in gathering enough light from a spiral nebula—in this case, the spiral in the constellation Andromeda—to measure its Doppler shift. The result was a shock to astronomers.

The Andromeda nebula was approaching Earth at a speed of 186 miles per second, at that time the highest speed ever measured for a celestial object. By 1914 Slipher had found the line-of-sight speeds for fifteen spiral nebulae, and to everyone's surprise, thirteen of them were receding from Earth, some almost 500 miles per second faster than Andromeda's speed of approach. For

1923-1924 Edwin Hubble found Cepheids in Andromeda, then used Shapley's work with Cepheids to calculate a distance for the nebula, establishing the existence of galaxies beyond the Milky Way.

Curtis and like-minded astronomers, this was a powerful argument for the island universe theory. The speeds seemed entirely too great for the spirals to be gravitationally bound to the Milky Way.

Shapley disagreed. Even if the spirals were not within the Milky Way, he pointed out, they could be just outside it and therefore not large systems of comparable size. He explained their high speeds and general recession from Earth as the result of the pressure of radiation from the galaxy (a theory that failed to stand up, given the much greater gravitational attraction the galaxy would exert). But Shapley had another reason for believing that the spiral nebulae were nearby. Since 1916 a colleague at Mount Wilson, the Dutch-born and -educated Adriaan van Maanen, had been finding what he believed were rotational motions for spirals. The work involved extremely exacting measurements, generally with the aid of a stereocomparator. Such a device can superimpose two star photographs taken at different times. An astronomer then measures the differences between the two photographs, such as changes in the position or brightness of a particular star.

Van Maanen used the stereocomparator at Mount Wilson to compare photographs of spiral nebulae taken several years apart. He was looking for movements within the arms, and after much painstaking work he reported that he had found them. The movements were very small, but they showed up in more than one spiral and seemed to be roughly consistent. Van Maanen even calculated the time that it takes a spiral nebula to make one complete revolution; the periods ranged from 60,000 to 240,000 years.

To Shapley these results settled the issue absolutely. If the spiral nebulae were distant galaxies comparable to the Milky Way, movements of the type van Maanen had observed would not be visible. Rotational motion in the spirals had in fact been detected spectroscopically by Vesto Slipher and others: The spectral lines from the receding side of the spirals were red-shifted; those from the approaching side were blue-shifted. But detecting rotation optically, as van Maanen had done, was akin to being able to measure the parallax of a star, in that it meant the spirals had to be nearby. To assume otherwise led to a logical absurdity: With a period of rotation of 240,000 years, the spiral arms of a distant galaxy comparable in size to Shapley's huge Milky Way would have to be moving faster than the speed of light.

Curtis admitted that van Maanen's findings were a blow to the island universe theory. Van Maanen was a respected scientist, and his procedures seemed unassailable. All Curtis could do was question the precision of his measurements and reserve judgment.

THE END OF THE DEBATE

In retrospect it is clear that both Shapley and Curtis were right about some things and wrong about others. At the time, the information available to them simply was not good enough for either man to carry the day. Even as the debaters were arguing their points, however, the astronomer who would settle the issue was at work back on Mount Wilson. Like Shapley, Edwin

Hubble was born in Missouri and came to astronomy after a stint in another profession. But in all other respects the two men were radically different. Hubble spent his undergraduate years at the University of Chicago, then traveled to England on a Rhodes scholarship to study law. Ever after, he spoke with an English accent, an affectation that irritated Shapley no end. When Hubble returned to the United States he began practicing law, but a boyhood fascination with astronomy eventually drew him back to the University of Chicago and its Yerkes Observatory in Williams Bay, Wisconsin.

Just as Hubble was finishing up a doctorate in his new field, Hale was getting ready to hire staff for the soon-to-be-completed 100-inch telescope on Mount Wilson, and he invited Hubble to be one of the astronomers to use it. But the year was 1917, and in reply to his offer, Hale received the following telegram: "Regret cannot accept your invitation. Am off to war." Hubble enlisted in the army and was sent to France. When he returned to the United States in 1919, Hale's offer was still open, and Hubble immediately accepted.

He began with some work aimed at classifying the nebulae *(pages 32-33)*, a subject he had investigated as a graduate student. This pursuit embraced what had always been his real interest: the spiral nebulae. Like Curtis, Hubble believed in the island universe theory, and he hoped that Mount Wilson's 100-inch would validate it. But even that powerful instrument could not unequivocally capture individual stars in photographs of the spirals. Hubble himself thought he could see stars, but other astronomers—including Shapley—were skeptical. In any case, simply finding stars would not resolve the issue if there was no way to determine how far away the spirals were.

Despite the uncertainty, the photographs of spirals were in many instances sharp enough to reveal the pinpricks of light that Curtis had identified as novae, and Hubble kept a sharp eye out for these. One day in 1923, when he was working through several photographic plates of the great nebula in Andromeda, he reexamined a spot of light that he had first taken to be a nova and had marked with a capital *N*. By tracking back through earlier plates, Hubble was able to plot a light curve and realized that the star brightened and dimmed periodically. In triumph, he crossed out the *N* and wrote "VAR!" next to the star. The erstwhile nova had behaved exactly like a Cepheid variable. Finally Hubble had a way to determine the distance to the Andromeda nebula. The Cepheid in the spiral was very dim, much dimmer than the ones Shapley had found in the globular clusters. Using Shapley's calibration, Hubble calculated that the Andromeda nebula was 900,000 light-years away—well beyond even Shapley's Milky Way. Andromeda was a full-fledged galaxy.

Always very cautious about publishing his findings, Hubble did not rush into print with his news. Perhaps because of his training in law, he was a methodical researcher, keen to erase any shadow of doubt before presenting his results in public. And there was doubt, in the form of van Maanen's supposed measurements of motions in the spirals—including Andromeda. Moreover, the period-luminosity relation was still controversial as a distance indicator. Finally, however, at the repeated urging of friends, he prepared a

In a journal of observations he made in 1923, Edwin Hubble marked his discovery of a variable star in the Andromeda nebula with an arrow to point out the relevant photographic plate. Later he showed the variable and its host nebula to be well beyond the edge of the Milky Way—the first convincing proof of the existence of other galaxies.

HUBBLE'S GALACTIC SPECIES

Edwin Hubble's effort to classify the "extragalactic nebulae," as he called galaxies, was a lifelong pursuit. The scheme shown here, published in 1936, is a revised version of a system he had proposed a decade earlier, based on his examination of hundreds of photographs made with Mount Wilson's 60- and 100-inch telescopes. As his collection of photographs grew, he modified the scheme but kept its essential features.

Hubble divided galaxies into two broad types: regulars, which show rotational symmetry about a central nucleus, and irregulars, which lack this symmetry (and thus were not included in his sequence). Regular galaxies were divided into ellipticals and spirals. Two branches of spirals—one with a bar across the nucleus—were classified according to size of the nuclear region and openness of the arms. Other schemes have since been developed, but Hubble's basic classifications are still the most widely used.

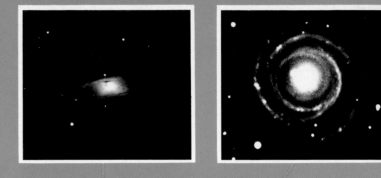

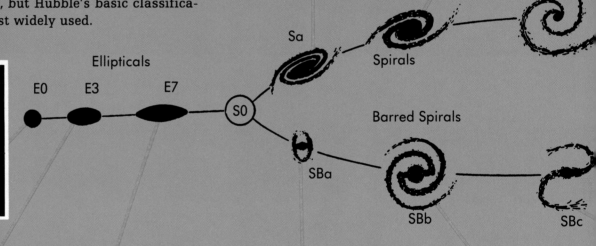

Ellipticals

E0   E3   E7

Sa

Sb

Sc

Spirals

S0

SBa

Barred Spirals

SBb   SBc

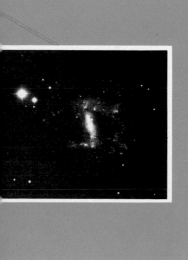

In Hubble's 1936 system, regular galaxies progress from featureless ellipticals to more complex spirals and barred spirals. The two spiral groups each have three subcategories—a, b, and c—representing a sequence toward smaller nuclear regions and more open arms. S0-type galaxies—which Hubble had not actually observed when he devised the diagram—act as a bridge between the ellipticals and the spirals. Flatter than an E7, the most elongated elliptical, S0s lack spiral structure. Hubble theorized—accurately—that the transition from E7s to Sa-type normal spirals might be less abrupt.

paper on his discovery. Henry Norris Russell, who had been Shapley's mentor at Princeton and was a leading authority on variable stars, presented it at a meeting in Washington, D.C., on January 1, 1925. As recently as ten months earlier, Russell had still accepted van Maanen's measurements as proof that the spirals lay within the Milky Way. But Russell was even more firmly committed to the reliability of the Cepheids as distance indicators. Faced with a choice between abandoning Cepheids and accepting spirals as extragalactic, Russell endorsed Hubble's work.

Finding Cepheids in the spirals marked the beginning of a new age in astronomy. With a single stroke, Hubble had increased the volume of the universe a hundredfold. Most astronomers immediately accepted the finding, although many, including Shapley, took a few months to be persuaded.

As for van Maanen's measurements of motions in the spirals, scholars have carefully scrutinized his procedures in a search for any kind of systematic error that could have produced his results. No such error has been found. The most likely explanation is that van Maanen found motions in the spirals because he expected to find them. He was working at the very limits of his instruments' precision, and a consistent error of a tiny fraction of a millimeter could have led him astray. Not until the mid-1930s, after Hubble had made a direct assault on the measurements, did van Maanen relent. "It is desirable to view the motions with reserve," he wrote in the *Astrophysical Journal* in 1935. The final objection to the island universe theory had been laid aside.

Hubble called his discovery "an achievement of the great telescopes," but there is more to it than that. Smaller telescopes could have found the Cepheids in the Andromeda galaxy—if anyone had thought to look for them. But other astronomers were too skeptical of Shapley's use of Cepheids to mount such a search. And Shapley, convinced that the spirals were not galaxies, had no reason to look. Still there is no denying that the 100-inch was an important factor in Hubble's discovery. With it he succeeded in resolving the outer portions of galaxies into stars (namely, the very bright Cepheids), an achievement that had eluded everyone until then. And it played an even larger role in his subsequent work. As Shapley had done before him, Hubble used the Cepheids to develop new kinds of distance indicators, but this time for galaxies rather than globular clusters.

He first compared the apparent magnitudes of the brightest stars in a long list of galaxies with the apparent magnitudes of the brightest stars in galaxies containing Cepheids, whose distances could therefore be estimated. The comparison allowed him to calibrate the absolute magnitudes of the bright stars in galaxies without Cepheids and thus the distances to those galaxies. This work also gave him a way to gauge the absolute magnitudes of entire galaxies. Then, at distances where individual stars could not be resolved, he compared the brightnesses of the galaxies themselves. In this fashion, he worked his way ever farther into the universe, using the galaxies as stepping-stones.

Meanwhile, Vesto Slipher at the Lowell Observatory had been continuing to measure Doppler shifts for spiral nebulae—subsequently recognized as spiral galaxies. By 1925 Slipher had speeds for forty-five of them, including many for which Hubble was determining distances. The torch then passed to Milton Humason, a thorough researcher who had become Hubble's assistant. Using the 100-inch telescope at Mount Wilson, he accumulated even more red-shift evidence indicating that most galaxies were receding from Earth, and at unheard-of speeds.

These results found support among theoretical physicists, whose work suggested a possible link between the red shifts and distances of astronomical objects. Hubble now focused on that connection. He compared the distances he was getting for the galaxies with the speeds Slipher and Humason were deriving. A simple relationship emerged: With the exception of nearby galaxies, the farther away a galaxy was, the faster it was receding. He determined the rate at which this recessional velocity changes with distance, a number now known as the Hubble constant.

THE EXPANDING UNIVERSE

When his results were published in 1929, they rocked astronomy to its foundations. Hubble's work showed that the universe was expanding; all of the galaxies were rushing away from each other, increasing the space between themselves and their neighbors. Perhaps more significant, the work could be applied to a problem that was beginning to intrigue astronomers and astrophysicists—namely, finding the age of the universe. Theoretically one could run the expansion scenario in reverse to a point when all of the galaxies would come together, thus dating the beginning of space and time.

One of the ironies of modern astronomy is that even though Shapley and Hubble were right in broad terms about, respectively, the scale of the Milky Way and the extragalactic nature of the spiral nebulae, the actual numbers they invoked to support their conclusions were quite crude. Because Shapley did not take into account factors such as the absorption of light by intervening dust, his estimate for the optical diameter of the Milky Way was three times too large. A more accurate value is about 100,000 light-years. Similarly, the value now accepted for the Hubble constant is between one-tenth and one-fifth of Hubble's original number. Much current work in astronomy is aimed at determining it more precisely or finding other ways to express the relationship. Even the exact calibration of the period-luminosity relation for Cepheid variables, the cornerstone of extragalactic measurements, remains uncertain until the distance to a Cepheid can be measured directly.

Thus, the age-old question *how far?* continues to haunt astronomy. As Hubble wrote in one of his last papers, "With increasing distance our knowledge fades, and fades rapidly, until at the last dim horizon we search among ghostly errors of observations for landmarks that are scarcely more substantial. The search will continue. The urge is older than history. It is not satisfied and it will not be suppressed."

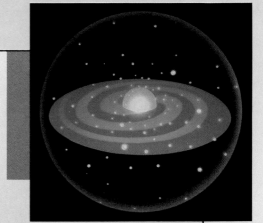

THE ARCHITECTURE OF A SPIRAL GALAXY

A mong the most beautiful objects in the universe are the galaxies that the astronomer Edwin Hubble classified as spirals—great pinwheels of starshine spinning in the void. Perhaps three-quarters of all galaxies are of this type, including our own galactic home. They vary considerably in size, ranging from giants like neighboring Andromeda *(page 4)* to compact cousins that may be only a tenth as big. In the finer details of their appearance, too, they display great diversity—their luminous arms forming complex tangles, their glowing gases and billions of stars tugged awry by encounters with other galaxies. But, as explained on the following pages, a common structure clearly distinguishes these star systems from ellipticals and other species in the galactic zoo. Spiral galaxies have a central bulge of stars, a large, pancakelike disk extending outward from this core, and a spherical stellar halo that encompasses both bulge and disk. They also show signs of possessing a feature that no earthly instrument has yet been able to detect: Around the visible parts of the star system seems to lie a so-called dark halo, an inert and mysterious envelope of matter whose existence can be inferred only from the galaxy's motion.

A CATHEDRAL OF LIGHT

Unlike elliptical galaxies—simple shapes that look like nothing so much as immense clouds of light—spiral galaxies are rich in structural detail. Their most distinctive visible components fall into two sub-systems: a bubblelike halo centered on a dense sphere of stars known as the bulge, and a thin, flat disk of stars and gas inscribed with whorls of alternating light and dark lanes.

The glowing halo, far dimmer than either the disk or the bulge, can be as much as 400,000 light-years in diameter and is defined by the orbits of individual stars and ball-shaped collections of stars called globular clusters. Most of the clusters travel within the sphere, but a few of them—known as intergalactic tramps—follow orbital paths that throw them hundreds of thousands of light-years beyond the halo's diffuse boundaries.

The galactic disk and the bulge at the center of the halo vary in proportion to each other from galaxy to galaxy. In some cases, the bulge spans 100,000 light-years, nearly swallowing the disk and its pattern of spiral arms. In other stellar systems, the disk is as much as 100,000 light-years across and the bulge a small protuberance embedded in the middle of it.

Typically, the disk has a thickness only about one one-hundredth its diameter. Within this narrow plane, the pattern of spiral arms rotates slowly about the galactic center. Few galaxies display the coherent two-arm structure of the stylized galaxy shown here and on the following pages. More commonly the arms of spiral galaxies are fairly fragmented and indistinct, with ragged spurs poking out from one arm or another. Beyond the visible portion of a galaxy lies yet another galactic component, made up of so-called dark matter, whose presence has been detected only indirectly *(pages 42-43).*

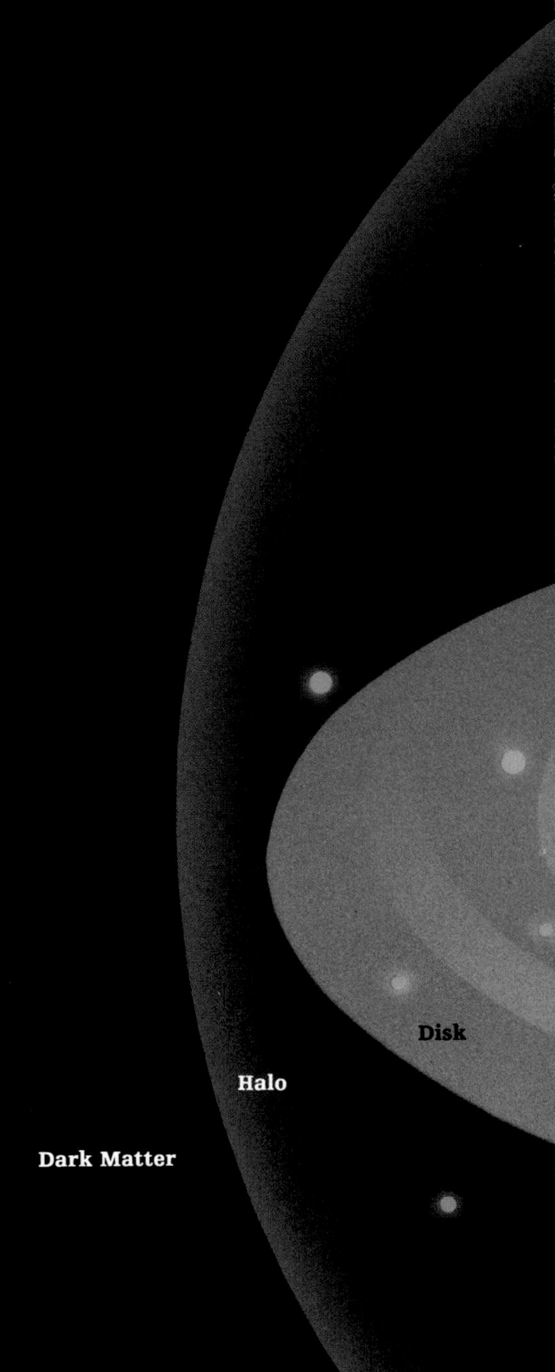

Disk

Halo

Dark Matter

Bulge

COMPONENTS OF THE BULGE AND DISK

The billion or so stars that inhabit the bulge of a typical spiral galaxy differ in a few characteristic ways from those in the disk. For one, bulge stars swirl around the galactic center in highly eccentric orbits, as indicated here. Steeply inclined to the plane of the disk, these orbits fling the stars out tens of thousands of light-years before bringing them back in near the nucleus. The orbits of disk stars, in contrast, are nearly circular and general-

ly lie within 300 light-years of the plane of the disk.

The two populations of stars also differ in their ages, and thus in color and luminosity. Astronomers estimate that bulge stars may be at least 10 billion years old, dating from very early in a galaxy's life. The disk, for its part, is an active stellar nursery, whose youngest members are only a few million years old. Unlike the bulge, which contains virtually no interstellar dust and gas, the disk is a vast reservoir of this star-making material. Through the billions of years of a galaxy's life, the birth of new stars—particularly of hot, massive blue stars that tend to shine very brightly—creates the luminous pattern for which spiral galaxies are named. Most disk stars live long enough to orbit the center of the galaxy tens or hundreds of times. But the heaviest—up to fifty times more massive than the Sun—die before they can complete even one orbit of the center.

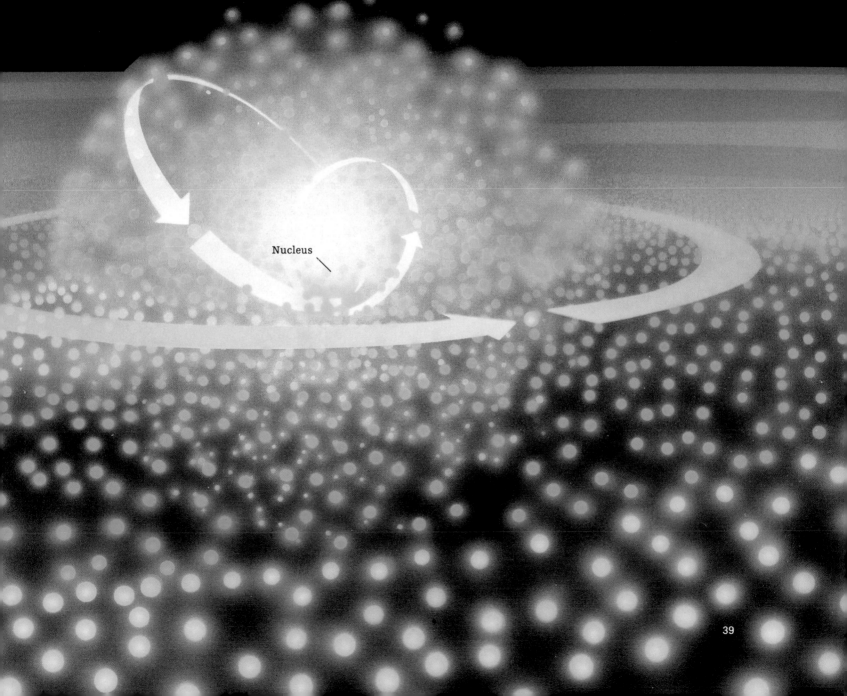

Nucleus

A Halo of Clusters

One of the first discoveries in the relatively short history of galactic astronomy was that spiral galaxies possess an extended halo defined by huge groupings of stars known as globular clusters. Ranging from 15 to 300 light-years in diameter, these clusters may contain anywhere from tens of thousands to a few million stars, a mass roughly one-millionth that of an average spiral galaxy. The stars are primarily cool, ancient red stars, similar to those in the galactic bulge.

Like the individual stars in the bulge, globular clusters follow extremely eccentric orbits around the galaxy's center, rising as much as 300,000 light-years from the galactic plane. Thus, twice on every several-million-year orbit, the clusters pass through the plane of the disk.

The passage spawns no spectacular stellar collisions: Both disk and cluster stars are much too widely dispersed for direct encounters to be likely. Interstellar gas is a different matter. Unlike stars, gas is susceptible to the effects of turbulence and viscosity. As a cluster dives through the disk's layer of gas, the layer interacts with—and draws off—the gas that has accumulated in the cluster. By repeatedly losing the material for making new stars, the clusters may, over the course of hundreds of billions of years, gradually transform themselves from collections of long-lived red stars into an assemblage of burned-out white dwarfs.

A Massive Darkness

One of the most perplexing aspects of galactic architecture is the featureless dark halo extending at least 100,000 light-years beyond the halo outlined by the globular clusters. It has never been directly observed, but several lines of argument, including optical studies of galactic rotation rates *(below)*, indicate that such a halo surrounds most typical spiral galaxies.

Although the gravitational effects of this dark matter can be measured, its identity is a mystery. Any candidate must lack detectable radiation yet possess enough mass to add up to the total necessary to have the observed effect on galactic rotation. Among the possibilities are exotic fundamental particles bearing such names as the magnetic monopole, the axion, and the photino; the equations of quantum physics seem to allow for them, but none of these phenomena has been reliably observed. Another contender—known to exist—is a particle called the neutrino, but astronomers are still uncertain of its mass. More substantial candidates—low-mass stars or black holes—cannot be accounted for in large enough numbers. Combinations of candidates might solve the puzzle, but most astronomers find such solutions unsatisfying, leaving the nature of dark matter an open question.

Solid body rotation. In a rotating solid disk, such as a phonograph record, the outer edge revolves faster than the inner portion.

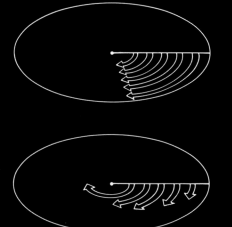

Solar System rotation. In the Solar System, where the Sun holds most of the mass, planets orbit more slowly the farther out they are. For example, Mercury, the closest planet, travels ten times faster than Pluto, which is a hundred times further from the Sun.

Galactic rotation. In a galaxy, mass is more widely distributed; rotation rates of gas and stars should increase with distance from the center until most of the galaxy's mass is inside their orbit, then slow farther out *(red)*. In fact, galactic rotation rates never drop *(white)*, evidence that unseen matter well beyond the visible disk controls the stars' velocities.

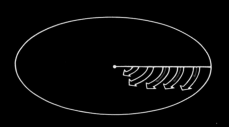

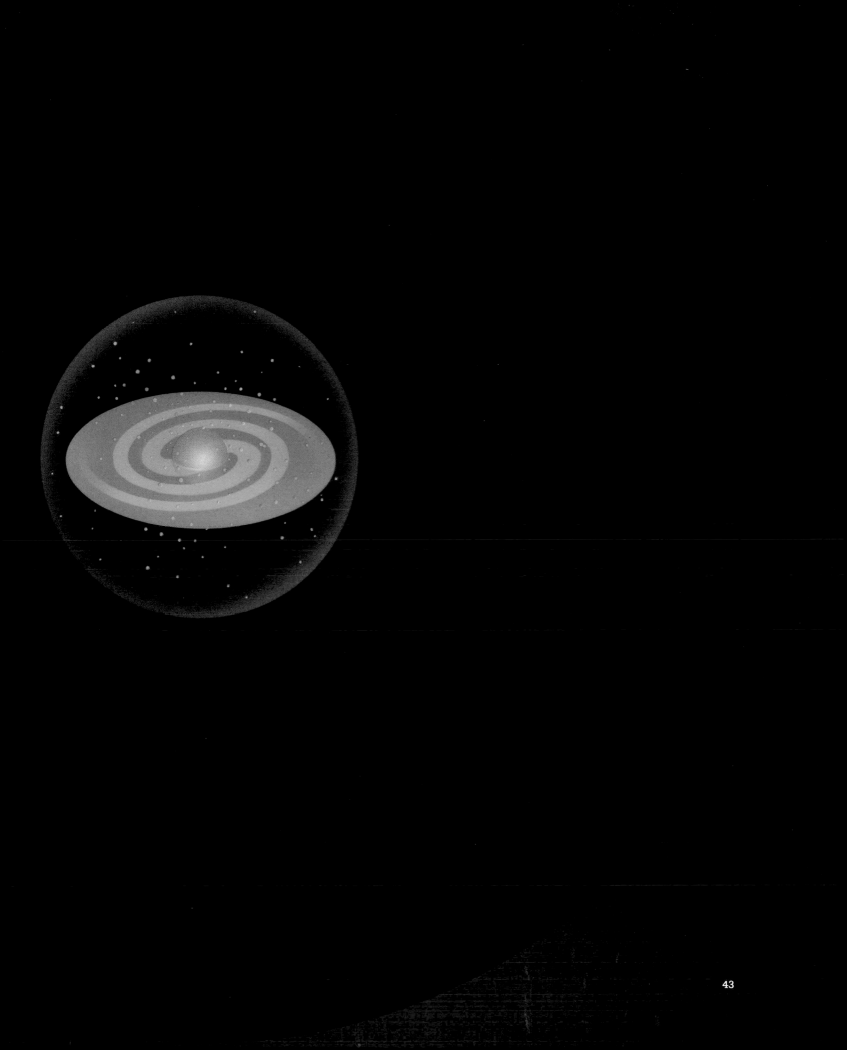

Blazing white clusters of stars
(lower left) lie in the constellation
Sagittarius, guidepost to the heart
of the Milky Way. Interstellar dust,
densest along the galactic plane,
reddens the light of gas clouds.

uring the summer of 1914, his first on Mount Wilson, Harlow Shapley paid a call on an eminent visitor staying at one of the observatory cottages. Dutch astronomer Jacobus Kapteyn, then in his early sixties and in frail health, was not exactly a stranger to the place. He had spent the better part of a decade on a project that required him to make regular pilgrimages to many of the world's astronomical centers. In 1906 Kapteyn had persuaded his international colleagues to participate in a mammoth undertaking: They would map the size and shape of the Milky Way galaxy by surveying the stars in 206 specific areas of the sky. Although Kapteyn's final analysis of the star counts would not be published until 1922 (the year he died), the astronomer had already made preliminary calculations indicating the galaxy to be about 30,000 light-years in diameter and 6,000 light-years thick. His results also placed the Sun near the center of the system.

Shapley had some preliminary work of his own that he thought Kapteyn might find relevant. The young astronomer had recently begun his study of globular star clusters and was turning up distances to the stellar groups of 30,000, 50,000, and even 100,000 light-years. These dramatically exceeded the bounds of Kapteyn's model—known as the Kapteyn Universe, since most astronomers still assumed that the Milky Way and the universe were one—and Shapley looked forward to hearing what his illustrious colleague would have to say about the estimates.

Kapteyn listened politely to Shapley's explanation of his work, to his description of the Cepheid variables he had used as distance gauges, to the computations and deductions that had followed. He looked at Shapley's paper, and then, as Shapley remembered many years later, "suggested that I check my observations again." Although the Dutch visitor clearly would not accept Shapley's figures, the encounter ended amicably. "He was kind about it," Shapley recalled, "because I was a nice young man and he was a nice old man."

At the time, Shapley himself essentially subscribed to Kapteyn's model for the dimensions of the Milky Way, and he reasoned that the extraordinary distances to the globular clusters were evidence that the globulars were separate galaxies. Within a few years, of course, as he accumulated more data and refined his distance estimates, Shapley would conclude not that the clusters were island universes but that the Milky Way was ten times larger than most astronomers believed it to be.

Even after Edwin Hubble established that the Milky Way was just one among many galaxies, astronomers remained quite uncertain of its nature. The galaxy's size and shape were not its only mysteries; the motions of its stars were also puzzling. For example, astronomers had believed for hundreds of years that stellar movements were essentially random. Then, just after the turn of the century, Kapteyn's extensive analysis of thousands of photographic plates revealed that all the stars in the galaxy seemed to be moving in one of two great streams: Three-fifths were traveling toward the constellation Orion; the other two-fifths flowed in the opposite direction, toward the constellation Scutum. These movements hinted at some sort of deep cosmic order, its mechanism and details unknown.

Astronomers of Kapteyn's generation were determined to ferret out the secrets of the vast stellar nation to which the Sun belongs, and they approached the challenge with the best techniques at hand. But the validity of those techniques hinged on one major assumption. Although astronomers accepted that light absorption took place in localized dark clouds seen in the Milky Way, they assumed that interstellar space was transparent. If they were wrong about this, their distance estimates would be far off the mark.

Kapteyn himself was keenly aware of the uncertainty. He worried particularly that the transparency assumption had led him into error when he positioned the Sun near the center of the universe—an oddly favored location that was reminiscent of the old view of Earth as the pivot of the heavens. Over the years he initiated a number of efforts to gauge interstellar light absorption, but nothing conclusive ever came of them. He was right to worry: Within a decade of his death, several discoveries would relegate Kapteyn's Sun-centered universe to a historical footnote. A telling blow to the underpinnings of the Kapteyn system came from one of his own students, a gifted astronomer by the name of Jan Hendrik Oort, who in time would do a great deal to unravel the riddles of the Milky Way.

SORTING OUT THE STARS
In 1922, the last year of Jacobus Kapteyn's life, Oort was a twenty-two-year-old graduate student working under the old master's direction at the University of Groningen in the Netherlands. His first major task was to take a close look at a recently noted and enigmatic phenomenon: Astronomers had observed that, relative to the motion of the Sun, a few stars seemed to move at about twice the speed of other stars in the Sun's neighborhood, like falcons dropping through a flock of lazy pigeons. Oort's analysis showed that these anomalous objects, known as high-velocity stars, all seemed to travel toward a particular quadrant of the sky, but he was at a loss to explain where they came from. According to the laws of physics, such velocities could not be maintained within the small Kapteyn system, so Oort suggested that the stars might be interlopers. But this theory raised as many questions as it tried to

answer, and although he defended it in his doctoral thesis of 1926, he was himself dissatisfied with the explanation.

Hints of a solution to the mystery already existed. Harlow Shapley, in postulating his Big Galaxy theory in 1918, had noted that the globular clusters were heavily concentrated in the direction of the constellation Sagittarius. His model of the clusters' orbital system placed the focus of their orbits at a point tens of thousands of light-years from Earth, and Shapley concluded that this point must be congruent with the real center of the galaxy. As Oort wrote many years later, "I did not doubt that the center of the entire galactic system must coincide with that of the globular clusters, and that it must have a mass very much larger than that of the Kapteyn system in order to prevent the high-velocity stars and the clusters from escaping altogether." For the moment, however, the relationship between Shapley's system of globular clusters and the Kapteyn Universe remained murky.

Then in 1925—the same year Hubble proved that the spiral nebulae were galaxies comparable to the Milky Way—a twenty-nine-year-old Swedish astronomer and theorist named Bertil Lindblad tried to reconcile the two systems. Lindblad had studied mathematics, physics, and astronomy at Uppsala University and, upon receiving his doctorate in 1920, had spent two years doing postdoctoral research in the United States. During this period, he visited the Lick, Harvard, and Mount Wilson observatories, where he was exposed to both sides of the vigorous debate over the size and structure of the galaxy. Inspired by Shapley's work, Lindblad began analyzing studies on the high-velocity stars. According to the calculations he published in 1925, the apparent motions of these stars could be explained as a natural consequence of rotation. He theorized that faster-moving stars would tend to follow circular orbits that were flattened to the plane of the galaxy. Relatively slow-moving stars would tend to travel in more elliptical and more steeply inclined orbits that took them out of the galactic plane; such a subsystem would tend to be spherical, rather like Shapley's globular clusters.

CELESTIAL ILLUSION
Lindblad suggested that, despite their name, the so-called high-velocity stars belonged to the latter group. They were on inclined elliptical orbits and moving slowly; they appeared to be speeding past because the Sun was actually overtaking them. Similarly, Kapteyn's mysterious star streams were the result of the motions of stars in the galactic plane on a variety of slightly elliptical orbits that crossed the Sun's path. From the Sun's vantage point, half the stars would be on the outbound portion of their ellipse, the other half on the inbound portion.

Initially, Lindblad's theory attracted little attention among astronomers, possibly because his landmark paper was, as Oort later described it, "written in an obscure way." But Oort himself struggled through the paper's mathematical thickets to conclude that Lindblad's hypothesis was "a beautiful explanation" of observations that had seemed inexplicable. Moreover, Oort soon found a way to confirm and expand the theory by carrying out further analysis of stellar motions.

The Dutch astronomer hypothesized that the galaxy rotates in a manner analogous to the Solar System *(pages 42-43)*. As a result, he expected to find that stars closer than the Sun to the galactic center—and therefore traveling at higher orbital velocities—would appear to overtake the Sun and leave it behind. In contrast, stars farther from the center, which should be traveling more slowly, would seem to move past in the opposite direction.

When he compiled the data for stars in the Sun's neighborhood, he found that the Doppler shifts of their spectral lines followed the pattern of differential rotation he had predicted. In a paper published in 1927, Oort applied established principles of gravitation and motion to the pattern and determined that the components of the galaxy were rotating about a distant, massive center whose location was in Sagittarius. Although Oort's placement of the galactic center was within two degrees of the center of Shapley's globular clusters, he disagreed with Shapley by a substantial margin on the question of the Sun's distance from the center: Oort calculated that the Sun was approximately 19,000 light-years from the galactic hub, a little more than a third of Shapley's estimate.

AN INVISIBLE SCREEN

By the end of the decade, it was clear that Kapteyn's Universe was no longer an adequate description of the universe—or even of the Milky Way. But the new model of a larger, rotating galaxy still could not explain all of Kapteyn's observations. He had calculated the density of stars in space and found that, in every direction, the number of stars dwindled with increasing distance from the Sun. If the galaxy was in fact as vast as Shapley, Lindblad, and Oort suggested, something had prevented Kapteyn from detecting light from the farthest stars.

In 1930 proof of the obscuring medium was finally obtained—by a man who initially failed to realize what he had accomplished. But indirection was almost a way of life with Robert Trumpler. The son of a prosperous Swiss businessman, he was expected by all his kin to enter the family firm. However, a short stint in the world of commerce proved such an unhappy experience that he was able to convince his father to let him pursue his real interest, astronomy. By 1919 the thirty-three-year-old Trumpler had joined the Lick Observatory in California, where he spent the rest of his productive career.

Trumpler was a firm believer in Kapteyn's model of the Milky Way, and at Lick he began studies designed to increase understanding of the galactic structure. Throughout the 1920s he worked on determining the distances of so-called open clusters—stellar aggregations in which the stars are fewer and more widely dispersed than they are in globular clusters. Unlike the globulars, which are observed above and below the plane of the galaxy, open clusters are found in large numbers in the galactic plane.

To determine the distances to these stellar groups, Trumpler employed

DUSTY DECEPTION AND CONCEALMENT

When astronomers scan the night sky with optical telescopes, they are peering through a galactic fog that dims the light of stars and distorts Earthly perceptions of stellar distances and distribution. The culprit is dust—or more specifically, microscopic grains of rock that are coated with ice. Interstellar dust is created as stars age and die. In spiral galaxies, it usually accumulates in the central, star-making plane of the galactic disk. The Milky Way, for example, probably resembles the spiral at right (designated NGC 4565), with a dust layer running the length of the disk, like icing in a sandwich cookie.

The dust is extremely tenuous: Individual dust grains are even smaller than a particle of smoke, and they are separated from one another by an average of more than a hundred yards. Nevertheless, over great distances this rarefied material has a significant obscuring effect, scattering and absorbing the starlight that passes through it.

Because the Sun lies very close to the central plane of the Milky Way, astronomers on Earth gaze through dust no matter where they look. As shown below, unsuspecting astronomers at the turn of the century were thus hampered in their efforts to gauge the galaxy's size and shape. Today, direct observation of the dust layer is possible by means of infrared telescopes, which detect the dust's low-level radiation. Along with radio telescopes, these instruments also let scientists pierce the galactic veil to glimpse the structure of the Milky Way.

A dust grain is about ten millionths of a centimeter in diameter and probably consists of a rocky core inside an icy mantle. The frozen coating may be water, ammonia, or methane, or a combination of the three.

The bun-shaped universe. Astronomers once believed the Sun resided near the center of a relatively small universe of stars *(below, left).* In reality *(right),* the dimming effect of a band of dust in the midplane of the Milky Way's thin disk blots the light of more distant stars.

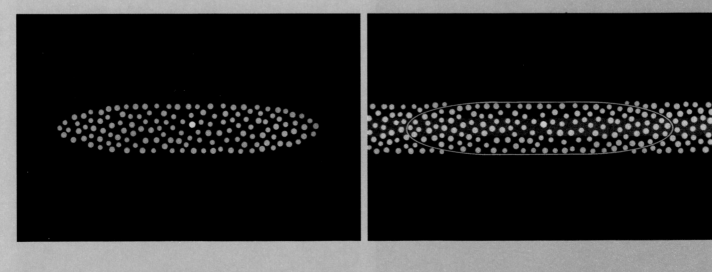

The zone of avoidance. Before they discovered interstellar dust, astronomers were puzzled by the way bright groups of stars known as globular clusters seemed to "avoid" the galactic plane *(below, left).* With instruments that penetrate the dust veil, the so-called zone of avoidance disappears *(right),* revealing not only more globular clusters but an enormous stellar bulge around the galaxy's center.

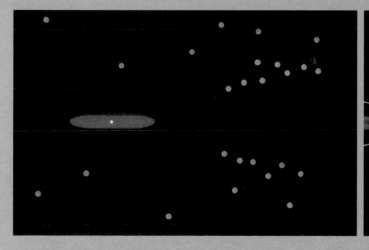

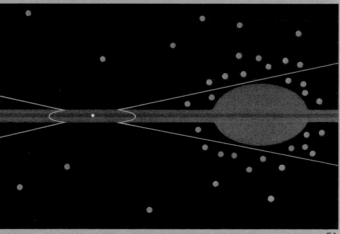

several methods of deducing the absolute magnitudes of individual stars in the clusters from their color and the lines in their optical spectra. He then compared the absolute magnitudes with the stars' observed brightnesses and calculated how far they were from Earth. (These methods are useful only for stars relatively nearby within the Milky Way.)

In June 1929 Trumpler reported that he had acquired enough distance information on open clusters to describe a Sun-centered system about 35,000 light-years in diameter—very similar to Kapteyn's Universe, in other words. As a check against possible errors, Trumpler next used the distances he had derived and the size of the clusters as seen on photographic plates to compute their actual diameters. What he found was utterly confounding: The farther away a cluster was, the larger its diameter.

Trumpler carefully went over his work, but he could find no errors. Only two possibilities remained, and one of them—that the clusters actually got bigger with distance—was unacceptable. There was no logical reason for it, just as, for example, there would be no logical reason for people to get taller the farther they lived from Philadelphia. This left the second explanation—that a thin layer of absorbing material exists throughout the galactic plane, dimming the light of stars in proportion to their distance from the Sun. The dimming effect would interfere with the measurement of their observed brightnesses and, consequently, the determination of their distances.

After performing additional calculations to test the second theory, Trumpler found that if light was dimmed by one magnitude for approximately every 5,000 light-years of distance, the anomalous relationship between distance and size was eliminated. In a paper published in 1930, he also suggested that the obscuring medium—later determined to be made up of tiny dust particles—must lie almost entirely in the galactic disk, so that objects such as globular clusters and galaxies lying above and below the plane of the disk are dimmed much less.

DECADES OF DETECTIVE WORK

Even as astronomers began to learn more about other star systems, their knowledge of the Milky Way remained frustratingly sparse. Buried in the plane of the galaxy and surrounded by light-obscuring interstellar dust, Earth was ill placed for a grand survey of the galactic whole. But over the course of several decades, resourceful investigators—some pictured at right and on the following pages—succeeded in piecing together the architecture of Earth's galactic home.

1901 Jacobus Kapteyn wrongly concluded from the distribution of known stars that the galaxy was 30,000 light-years in diameter and centered near the Sun.

1918 Harlow Shapley declared—also erroneously—that the Sun lay toward the edge of a galaxy ten times larger than Kapteyn had projected in his model.

Perhaps because of his bias in favor of the Kapteyn Universe, Trumpler did not immediately grasp the full implications of his discovery, continuing to espouse Kapteyn's relatively small, heliocentric system. But the effect of his work on the debate over the dimensions of the galaxy was profound. Harlow Shapley had assumed no absorption and thus had placed the globular clusters too far away. The 300,000 light-year diameter of his Big Galaxy was roughly triple the Milky Way's probable span (although that figure is still not precisely known). Kapteyn, in contrast, had adopted the absorption-imposed boundaries of observation as the limits of his Universe: Like a man standing on a long, fog-shrouded boulevard, he had counted the visible streetlights and concluded that he was on a very short street.

MAPPING THE SPIRAL STRUCTURE

Although Trumpler's work went a long way toward reconciling the conflicting measures for the dimensions of the galaxy, further progress in solving the many riddles of the Milky Way was derailed by the outbreak of World War II. Even among astronomers who managed to continue their studies, normal intellectual discourse was a rarity. But one researcher turned the exigencies of wartime to his professional advantage.

In 1931 German-born astronomer Walter Baade had joined the exodus of prominent scientists from his country. He made his way to California's Mount Wilson Observatory, a place he had first visited while on a Rockefeller fellowship in the 1920s. When the war began, many of his colleagues at Mount

1927 Jan Oort supported Lindblad's model of a large, rotating galaxy with a detailed study of star motions; his work confirmed the Sun's off-center galactic position.

1930 Robert Trumpler discovered a layer of obscuring dust in the galactic plane, thereby reconciling Kapteyn's small galaxy with Shapley's and Oort's larger systems.

1925 Bertil Lindblad explained the puzzling speeds and directions of certain stars with a theory that stars rotate at different rates around a common, distant galactic center.

Wilson left to work on projects of a military nature, but Baade was classified as an enemy alien and restricted to the immediate area around Pasadena and the observatory. The restriction—and the absence of competition—gave him virtually unlimited time on the observatory's 100-inch telescope.

Baade took full advantage of the opportunity. Wartime invasion jitters had led to a blackout of Los Angeles, and observing conditions could not have been better, for the 100-inch telescope no longer had to peer through skies polluted by glare from the expanding metropolis. Baade decided to concentrate on a challenge that had stymied astronomers for nearly two decades: resolving the central part of the Andromeda galaxy—that is, obtaining photographs that would show individual stars. Edwin Hubble, using the 100-inch in 1925, had been able to pick out bright stars in the galaxy's spiral arms, but the central area had remained a featureless blur of light. Now, in the fall of 1942, Baade noticed that photographic plates of the Andromeda galaxy showed, as he wrote two years later, "unmistakable signs of incipient resolution in the hitherto apparently amorphous central region." He realized that with the right observational techniques, a very small improvement in the telescope's resolving power "would bring out the brightest stars in large numbers."

Because Hubble had found blue stars in the spiral arms, Baade began photographing the central region with blue-sensitive plates. "It was natural to expect that the brightest stars in the central area, too, were blue," he would recall. "But I couldn't find any, and it was only after all the schemes for spotting blue stars had been exhausted that I turned to the red-sensitive plates to see if, perchance, the brightest stars were red."

1944 Hendrik van de Hulst predicted that neutral hydrogen in the galaxy's spiral arms could be detected by radio instruments tuned to the twenty-one-centimeter wavelength.

1932 Karl Jansky *(below)* inadvertently detected radio waves from space. Reading of Jansky's work, amateur Grote Reber *(right)* began charting the Milky Way's radio emissions in 1937.

1943 Walter Baade succeeded in finding individual stars in the center of the Andromeda galaxy. He discerned that the spiral arms contained very bright stars and gaseous nebulae.

As it happened, the observatory had recently acquired a new type of photographic plate that was extremely sensitive to red light. The plates had another advantage: They would allow longer exposure times. In theory, the longer the plate is exposed, the more light it can capture and the better the resolution of the image. Blue-sensitive plates become fogged after about ninety minutes because of high-altitude auroral light, a fluorescent, greenish glow caused when charged particles ejected by the Sun enter the Earth's magnetic field. With the red-sensitive plates and a special red filter, however, Baade expected to be able to expose the plates for up to nine hours. This was no trivial feat, since the slightest slip in the telescope's focus would blur the stars and blend them into the general background illumination. He determined that for the duration of the exposure he could not allow the focal plane of the huge 100-inch telescope to vary by more than a tenth of a millimeter—even though temperature fluctuations in the unheated observatory could warp the telescope's mirrors and cause changes of as much as several millimeters.

Baade tackled this problem in two ways. First, he chose the early fall for

1951 Three teams of radio astronomers—including, from left, J. V. Hindman, Harold Ewen, Alex Muller, and W. N. Christiansen—found van de Hulst's predicted twenty-one-centimeter emission line.

1966 Chia Lin *(left)* and his student Frank Shu *(below)* calculated that a spiral density wave could preserve the distinctive structure of the Milky Way and other spiral galaxies.

1951 William Morgan and other astronomers sketched an optical map of the Milky Way galaxy's spiral arms by locating bright objects of the type noted by Walter Baade.

making the observations, a season when nighttime temperatures on Mount Wilson remained almost constant. Then, on the day of an observation, he would open the observatory dome soon after lunch to let the temperature of the telescope's mirrors come to equilibrium with the outside air by nighttime. Second, he found a way to catch incipient focus problems very quickly: Instead of choosing a guide star that was centered in the telescope's visual field, he chose one that was somewhat off-center so that its image would be slightly distorted. Thus, instead of looking like a pinpoint of light—whose sharpness would be difficult to judge—the selected guide star resembled the fuzzy head of a comet, an elongated shape that tended to vary more obviously with changes in the telescope's focus. Baade set up the guiding eyepiece in such a way that very small changes in focus produced marked changes in the shape of the guide star.

A PATTERN OF POPULATIONS

All his precautions paid off in August and September of 1943. As he told the story later, "On the new photographs hundreds of tiny star images showed where none had been distinguishable before. They were red giants. As I turned the telescope to other regions of the Nebula, moving from the center outward, the pattern of prominent stars changed from red giants between the spiral arms to blue supergiants in the arms. It looked as though the central area and the regions between the arms were populated by one kind of star, whereas another kind predominated in the arms themselves."

Suddenly the stellar dynamics of distant galaxies were brought into high relief. To the bright, bluish stars concentrated in the spiral arms, Baade gave the name Population I. The less-luminous, reddish stars, which also occupied Andromeda's bulge and globular clusters, were called Population II. As other astronomers turned to analyzing stellar populations, it became clear that Baade's original scheme was too simple. By the late 1950s, researchers had come up with five population types, and in the decades since, as astronomers have discovered an ever-increasing variety of stars, the picture has grown even more complex. But Baade's two-population breakthrough had enormous practical applications for mapping the structure of the Milky Way. In 1950, during a symposium at the University of Michigan's observatory, he suggested that the supergiants of Population I, prevalent in the spiral arms of Andromeda, could serve as tracers to the arms in the Milky Way. A first step would be to study their spatial distribution in the solar neighborhood.

Sitting in the audience when Baade presented this mapping approach was William Morgan, a forty-four-year-old professor of astronomy at the University of Chicago's Yerkes Observatory. In the early 1940s, Morgan and two colleagues had refined a scheme that classified stars according to their spectral type. At the time of Baade's speech, he and his students were studying the distribution of bright stars and had developed new methods for deriving the distance to such stars through spectroscopic classification. The Yerkes team was well set up to put Baade's suggestion to the test.

Morgan and his colleagues concentrated on objects known as emission nebulae—vast, luminous gas clouds, such as the Orion nebula, with hot, young Population I stars embedded in them. Within two years, the team had its results. The nebulae seemed to outline two distinct spiral arms—one passing through the neighborhood of the Sun and a second lying about 6,500 light-years from the Sun in the direction of the constellation Perseus and away from the galactic center. A third arm seemed possible, located between the Sun and the galactic center in the constellation Sagittarius. The galactic map was still far from complete, but these findings offered the first solid evidence that the Milky Way was indeed a spiral galaxy.

Morgan and his team had capped three decades of rapid optical progress in the understanding of the Milky Way, but further insights into the galaxy's structure were still to come. At almost the same time that Morgan's group was completing its studies, a new application of a familiar technology—radio—would open up vast new areas of the sky that had been opaque to even the most powerful optical telescopes.

STELLAR STATIC

Radio astronomy actually began in the early 1930s at Bell Telephone Laboratories in New Jersey, where a young radio engineer, Karl Jansky, was

With this ninety-five-foot-long wood-and-piping contraption, Bell Telephone Laboratories engineer Karl Jansky *(below)* accidentally invented radio astronomy in 1932 while searching for the cause of static affecting overseas phone lines. By studying records of the radio noise received by the rotating device, Jansky could determine the direction of the source. He traced most of the outbursts to thunderstorms but also identified an extraterrestrial source—the center of the Milky Way.

investigating the sources of static affecting shortwave radio communication. Jansky detected one source that appeared with daily regularity but did not seem to be associated with any earthly transmitter. In fact, it appeared to travel with the stars, and it lay in the direction of Sagittarius.

The astronomical community paid little heed to Jansky's discovery, but in the Chicago suburb of Wheaton, Illinois, a twenty-five-year-old electronic engineer named Grote Reber was inspired by it. Starting in June 1937, Reber built himself what was, in effect, the first astronomical radio instrument. The device—a parabolic dish nearly thirty feet in diameter—weighed almost two tons and rested on a framework in Reber's backyard that raised it two stories high. Every morning and evening, as the rising and setting sun caused the metal strips that made up the dish to expand and contract, the neighborhood would be treated to a symphony of snapping, popping, and banging. During rainstorms, vast quantities of water would pour through the center hole of the dish, giving rise to rumors that the contraption was for collecting water and controlling the weather.

Over the next several years, Reber published articles describing his findings with this curious instrument in both radio engineering and astrophysical journals. By 1944 he had completed a map of the radio signals from the Milky Way; he even theorized that they came from hot interstellar gas.

Because most astronomers still regarded radio investigations as peripheral at best, Reber's work received little attention. But copies of his early articles found their way to the desk of Jan Oort in the Netherlands, despite the fact that the tiny nation was under Nazi occupation. By this time, Oort had inherited the mantle of his old mentor, Kapteyn: He was widely recognized as the leader of the Dutch astronomical community.

A skater who habitually frequented what one of his students described as "the thinnest ice on earth," Oort was equally daring in pursuit of intellectual quarry and especially alert to techniques that could open up new lines of research. But his adventurous self was not always in plain view. Many years later, another of his students, Hendrik van de Hulst, described a different persona. "Oort takes a scientific problem like hot food," van de Hulst recalled, "one bite at a time, lets it cool down, chews it, swallows it and digests it, and only much later tells how it tastes." In the spring of 1944, Grote Reber's work was the morsel of choice. As van de Hulst remembered it, Oort conveyed the significance of the young American's findings in an almost offhand manner, suggesting that van de Hulst look for a way to use this new technology to map the distribution in the galaxy of so-called neutral hydrogen—atoms of hydrogen consisting of a proton balanced by an electron and thus with no electrical charge.

In the early part of the century, astronomers had deduced the presence of rarefied gas in interstellar space from the spectra of certain stars. A few lines in the spectra did not display the Doppler shift characteristic of the other spectral lines but instead seemed to remain at rest. These stationary lines represented light-absorbing elements somewhere between a given star and

Earth. But the study of interstellar gases by optical means was limited by Earth's location in the dusty plane of the galaxy.

Because hydrogen is the most prevalent detectable component of interstellar matter, Oort thought it a likely candidate for radio investigation. He pointed out that if van de Hulst could find a radio emission line for neutral hydrogen, they could apply the Doppler shift method (pages 22-23) to the radio end of the spectrum and gain access to realms of the galaxy heretofore invisible from the dusty solar neighborhood.

FROM HINT TO BREAKTHROUGH

Van de Hulst turned to the problem of figuring out what might cause neutral hydrogen to emit a spectral line at radio frequencies. After detailed calculations and a couple of false starts, he settled on an event known as the spin-flip transition. In the course of its life, a hydrogen atom in space may occupy various energy levels. Van de Hulst was concerned with only two closely related levels: At one level, the atom's electron spins in the opposite direction to its proton; at the other level, it spins in the same direction.

Relatively common changes from one level to the other and back again—which result in changes of spin direction—occur when the electron collides with other atoms or stray electrons. Much more rare is a spontaneous change from the same to the opposite direction of spin. It is this spontaneous transition that gives rise to an emission of energy, detectable as a radio signal at a wavelength of just over twenty-one centimeters. Although an individual hydrogen atom might undergo this change only once every 10 million years, van de Hulst figured that the enormous quantities of interstellar hydrogen in the galaxy should produce a collective signal strong enough to be detected.

When van de Hulst announced his conclusions in 1944, Europe was still under siege, and the pursuit of ephemeral signals from space was not high on anyone's list of priorities. In any case, no instruments existed for searching out faint twenty-one-centimeter radiation. But as soon as peace restored some semblance of normality to the scientific world, Oort began putting together a team to go after the twenty-one-centimeter line. Oort's group acquired German radar antennas and set about building their own receivers. In 1951, however, a fire destroyed key pieces of equipment before their effort could bear fruit.

In the meantime, researchers in the United States had finally recognized the potential of radio astronomy and were also on the track of the twenty-one-centimeter line. The first to strike pay dirt wa Harvard graduate student Harold Ewen, working under physicist Edward Purcell. Their signal catcher was a crude, bulky horn antenna fashioned of plywood and copper foil. Mounted outside a laboratory window, the horn captured the hydrogen reading on March 25, 1951.

One of the ironies of the story is that van de Hulst himself was visiting Harvard during this period. When the line was recorded, Ewen generously gave him a description of the Harvard receiver, which van de Hulst imme-

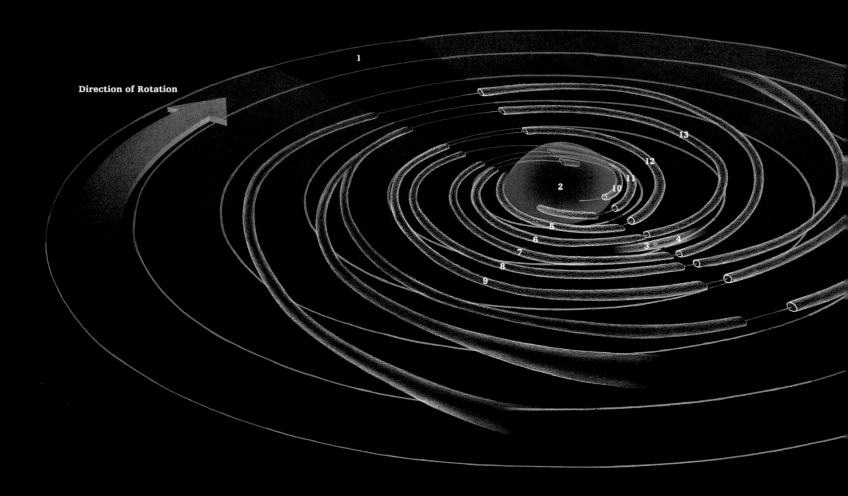

Direction of Rotation

A PORTRAIT OF THE MILKY WAY

Although astronomers have long believed that the Milky Way is a spiral galaxy, the case has been hard to prove from the Earth's galactic position. The first clear evidence came in the late 1940s, when bright, young stars were found in other spiral galaxies. Concentrations of such stars plotted near Earth seemed to trace long shapes that might be arm segments. At greater distances, however, interstellar dust hid the galaxy's secrets—until radio astronomy managed to penetrate the haze in the 1950s.

For radio astronomers, the sign of arms is the presence of clouds of neutral hydrogen and carbon monoxide, the stuff from which bright, young stars are made. The gas emits radio energy, and the red or blue shift of the radio waves indicates the distance and relative location of the clouds. Today, investigators have named nine bands of stars and gas. (Though named individually, each band is assumed to be an arm segment.) In two wedge-shaped regions of the galaxy, radio mapping has proved impossible: The contents of these regions are on tracks roughly parallel to that of the Sun and thus have almost no motion toward or away from Earth—the basis for redshift measurements.

Astronomers who produce models of the Milky Way differ over where to place the arms; they also connect the segments in the unmapped regions to form either two or four spiral arms. In the model shown here (placed over a light-blue grid with intervals of 15,000 light-years), the unmapped regions are shaded for clarity, and the segments make four arms. The Sun is located within a protrusion known as the Orion spur. Segments seen from the Northern Hemisphere are on the left side of the model; the Southern Hemisphere view is on the right. Recently observed segments, still unnamed, fill the outer part of the disk. Vital statistics of the Milky Way—estimates for the most part—appear in the box at right.

A Galactic Scorecard
Age: 13-15 billion years
Observed mass: 100 billion
  solar masses
Number of stars: 100 billion
Diameter of disk: 100,000
  light-years
Thickness of disk at Sun:
  700 light-years
Diameter of central bulge:
  30,000 light-years
Distance of Sun from center:
  27,000 light-years
Orbital velocity of Sun around
  center: 135 miles per second
Time for Sun to complete one
  orbit: 250 million years

diately relayed to Oort and electrical engineer Alex Muller back in the Netherlands. On May 11 the Harvard findings were confirmed by the Dutch, followed in July by a team of Australian astronomers. Later that year, in a fine show of international scientific solidarity, the results of all three teams were published together in the prestigious scientific journal *Nature.*

As Oort had hoped, finding the twenty-one-centimeter line opened a new era of galactic astronomy. Because radio waves allowed astronomers not only to "see" hydrogen directly but to penetrate interstellar dust, they could draw a much clearer picture of the galaxy. In fact, the dense concentration of neutral hydrogen in the galaxy's arms allowed it to serve—like Walter Baade's Population I stars—as a tracer to the spiral pattern. By pointing a radio telescope along a certain line of sight from Earth, astronomers could obtain a profile of the spectral line around the twenty-one-centimeter wavelength, with peaks corresponding to different concentrations of the gas at different velocities. By analyzing the Doppler shifts of each peak, scientists could figure out the relative location and distance of these concentrations, gradually building up a picture of individual arms or fragments of arms.

Although the Harvard team made the first discovery, subsequent efforts to survey the Milky Way with radio were led by the Dutch and Australian groups. Oort and his co-workers in the Netherlands were particularly diligent. Using two small hand cranks, they laboriously changed both the elevation and direction of their war-surplus antenna every two and a half minutes for nearly two years. The data they obtained for the Northern Hemisphere sky, combined with the Australian observations for the Southern Hemisphere, produced the first radio map of the galaxy. Known as the Leiden-Sydney map, it described what seemed to be four extended spiral arms linked by many branched structures.

WHY A SPIRAL?

By the beginning of the 1960s, astronomers were confronted with a Milky Way that appeared to be older, larger (except for Shapley's version), and far more complex than the one they had hypothesized forty years earlier. They knew that the thin central plane of the Milky Way contained enormous quantities of dust that obscured their optical view of the galactic hub, and that most of this dust, along with vast amounts of gas, was distributed among spiral arms, the birthing ground of new stars. With these facts in hand, they turned to the fundamental question of why spiral galaxies look the way they do. Why should there be spiral arms at all?

Bertil Lindblad, the Swedish astronomer whose theoretical work led to Oort's discovery of differential rotation, had spoken to that very issue more than three decades earlier. In 1927 he had been made director of the Stockholm Observatory, but theoretical puzzles continued to engage him. In trying to understand how arms could form and then how they were maintained, Lindblad devoted considerable time to painstakingly plotting the orbits of individual stars around the galactic center. He found that as a star travels in its

Fluorescent gas clouds and young blue stars brighten two arms curving out from the center of spiral galaxy M83.

Differential rotation can both create and destroy spiral patterns, as shown schematically at right with stars on either side of the galactic center *(1)*. As they orbit clockwise, the innermost stars finish half their circuits in the time it takes the outer stars to cover an eighth of theirs *(2)*, drawing the line of stars into two trailing curves. But as the stars continue to orbit the center, they draw farther apart relative to their starting positions, and the spiral arms stretch *(3)*. After the inner stars have completed one and a half orbits *(4)*, the arm structure is beginning to "wind up" about the center, an effect that becomes more pronounced over time *(box)*.

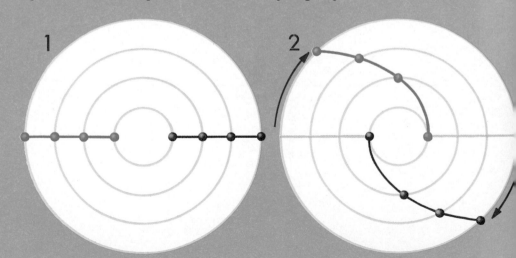

THE SPIRAL RIDDLE

Spiral galaxies represent three-quarters of the galactic population in the known universe, and they have long been the subject of intense scrutiny. Yet these cosmic wheels of light continue to present astronomers with a tantalizing conundrum: By all rights, the spiral arms that are their most striking characteristic simply should not exist in such great numbers.

At first glance, spiral arms might seem to be the natural result of a galaxy's spin—rather like the swirl of cream stirred into a cup of coffee. Studies of the motions of the stars, dust, and glowing gas that make up the arms show rates of rotation that vary according to the distance from the galactic center: The innermost stars and gas clouds in the arms may complete about half a dozen orbits in the several hundred million years it takes their counterparts near the galaxy's rim to cover their longer orbits once. This differential rotation, as it is known, can create a spiral structure. But in relatively few rotations, the pattern should smear *(below)*, just as well-stirred coffee quickly turns a uniform brown.

The arm structure should therefore be ephemeral by cosmic standards, particularly since the bright newborn stars that define the arms are known to be short-lived blue giants. At an average of twenty times the mass of the Sun, these stars consume hydrogen at such a breakneck pace that most collapse and die in less than 10 million years. (The Sun, in contrast, has already lived about 500 times that long.)

Hence the paradox: In defiance of the laws of probability, the cosmos is rife with spiral galaxies, most estimated to be at least 10 billion years old, sporting reasonably distinct—and yet presumably temporary—spiral patterns at the same moment in time.

Since the mid-1960s astronomers have attempted to find a mechanism that explains this unlikely phenomenon. The most widely accepted possibility is described on the following pages.

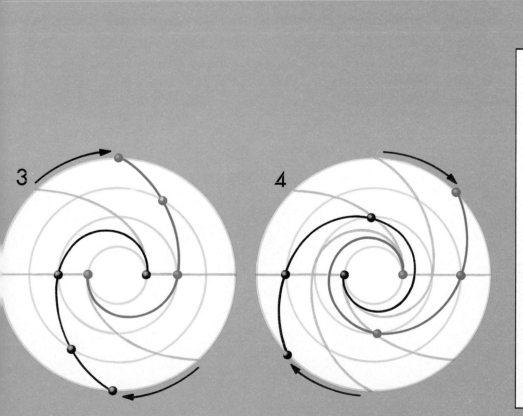

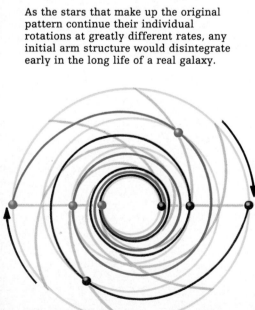

As the stars that make up the original pattern continue their individual rotations at greatly different rates, any initial arm structure would disintegrate early in the long life of a real galaxy.

A Self-Renewing Pattern

Because of differential rotation, any temporary spiral structure in a galaxy should disappear in the course of roughly a billion years. Illustrated here is a theory that explains how the spiral feature persists. According to this hypothesis, the arms observed in a galaxy at one time *(left)* are made up of entirely different stars and gas than the arms of the same galaxy several million years later *(right)*. The proposed mechanism for generating new arms is called a density wave. (Density wave theory does not by itself explain all types of spiral formations, nor does it address the question of how density waves originate.)

By definition, a wave is the movement of a disturbance through a medium; it is not the movement of material. For example, as a ship plows through water,

As a spiral-shaped wave of density *(white)* moves slowly around a galaxy, orbiting stars and gas trace a circular path *(pink)* that periodically intersects the wave pattern, causing new stars to form *(box and below)*.

In an enlarged view of the boxed area *(above)*, an orbiting hydrogen gas cloud *(pink)* enters the dense region of a spiral arm *(white)* along with a few stars. The region's greater mass exerts a strong gravitational force that slows the stars, causing the arm's density to increase until the stars eventually pass through. Because the gas cloud is more diffuse, the arm's greater density acts to slow the cloud down much more forcibly.

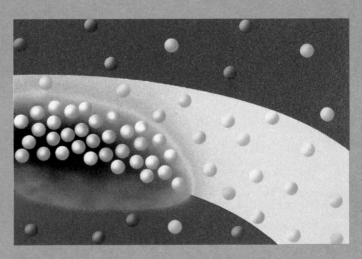

Almost as though hitting a wall, the gas in the hydrogen cloud rapidly compresses. Its original density is increased by a factor of ten, triggering the formation of a cluster of new stars *(white)*.

it disturbs nearby water molecules, which in turn bump molecules next to them, and so on. A wave pattern travels across the water, but no water actually travels with it.

Similarly, a galactic density wave is thought to be a spiral-shaped disturbance that moves through the galactic disk. The galaxy's stars, gas, and dust orbit the galactic center at varying rates, but all travel faster than the wave pattern itself. Thus, they periodically catch up with—and fall into—the slower-moving spiral, bunching up, much as a knot of cars might form around a large, slow truck on the highway.

The greater gravitational attraction of the dense arm region slows down the rotating stars, temporarily adding to the arm's density. The stars, with mass and momentum of their own, eventually pass through the dense region. Large, diffuse clouds of hydrogen gas, in contrast, tend to accumulate until forces of compression cause a burst of star formation *(below)*—in effect, turning on the spiral's lights.

Millions of years after the view at far left, the density wave has rotated to a new position, forming a new spiral arm. Stars that made up the earlier arms *(dashed line)* have dimmed or dispersed.

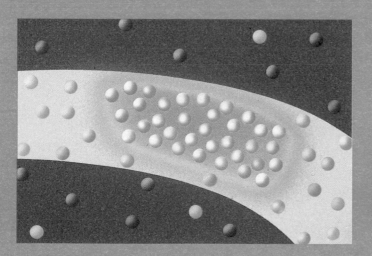

A cluster of young stars, some very massive and therefore very bright, passes through the dense region. These massive stars consume their fuel at such a rapid clip that they will die before they make it all the way through. But their ephemeral brilliance lights up the region, creating the spiral patterns visible from Earth.

As the remaining stars in the original cluster continue on their orbits around the galaxy, another gas cloud *(pink)* approaches the region of density, where the shock of compression will give rise to a new cluster of stars. The constant movement of stars and gas through the region causes the spiral pattern itself to shift gradually on a stately orbit around the galactic center.

elliptical orbit, gravitational perturbations from other stars cause the long axis of the ellipse itself to rotate around the galactic center, producing a complex orbital path. When many of these patterns are plotted and super-imposed on one another, they create a distinct spiral structure. Lindblad found that the structure rotates at a rate independent of the motions of the stars themselves. Borrowing a concept from physics, he concluded that the arms of spiral galaxies are the products of density waves—areas where fast-moving stars are temporarily clumped.

RESURRECTION OF AN IDEA

Lindblad published his theory on spiral arm development in the late 1920s, but it did not begin to gain wide acceptance until the mid-1960s. At that point, the concept was refined by two Chinese-American astronomers at the Massachusetts Institute of Technology, Chia Lin and his student Frank Shu. The Lin-Shu density wave theory *(pages 64-65)*, while adequate in certain respects, is by no means a perfect, all-inclusive explanation of the observed universe. As a theoretical model, it generates well-defined, two-arm spirals—not uncommon in the cosmos—but fails to account for the messy structure of the great majority of spirals, including the Milky Way. The theory also falls short on the origin issue. Although density waves provide a mechanism for the maintenance of the spiral arms over time, the Lin-Shu theory simply presupposes the existence of regions of density without addressing how they came to be. One possible cause is close encounters between galaxies: The enormous gravitational forces involved in a near-collision could disturb a galaxy's rotation enough to trigger the formation of such regions.

Another promising model for both the origin and maintenance of spiral arms is the random star-birth theory, which predicts the continual birth and death of spiral segments. Theorists suggest that the organizing force in the arms may be provided by local supernovae, whose explosions send out powerful shock waves that could compress nearby gas and dust clouds and trigger the formation of stars. Presumably, the differential rotation of the galaxy would draw the newly formed stars into the filaments we see as arms. In the late 1970s, computer models of the theory generated long-lasting spiral structures that resembled the messy spirals observed throughout the universe. However, the model does not generate the simple two-armed version produced by the density wave theory. Scientists thus are left with several possibilities: One method or another may operate in a given galaxy, or spiral arms may be formed by some sort of combination of galactic near-collisions, density waves, and random star births.

THE HEART OF THE MATTER

Even as some researchers struggled to explain the patterns of the Milky Way's disk, others turned their attention to the galaxy's hidden core. Discovered as a source of radio waves by Karl Jansky in the early 1930s, the galactic center had received relatively little attention in the next two decades. Then in 1957

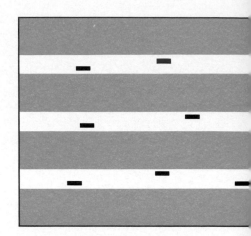

A density wave, similar to those thought to create the arms of spiral galaxies, forms around a slow-moving truck on a highway. Like a speeding star, a fast-moving red car approaches the knot of traffic waiting to pass the truck *(top)*. Entering the clump, the car slows to the speed of the truck *(center)*; once past, it resumes speed as traffic thins out *(bottom)*. The apparently unified mass of cars, its makeup constantly changing, constitutes a density wave moving slowly down the road.

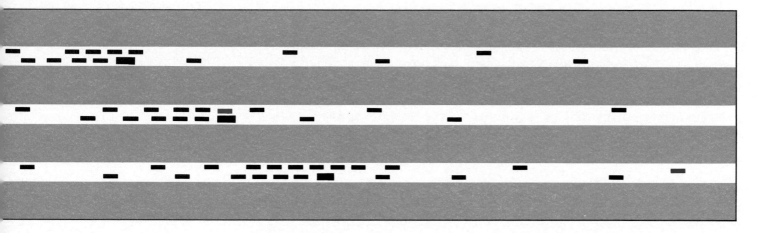

Jan Oort and a colleague, observing at twenty-one centimeters, found the first of two immense, arclike structures of neutral hydrogen expanding in opposite directions on either side of the nucleus. (The second, on the other side of the galaxy from Earth, was found several years later.) These features lie in the galactic plane rather like fragments of arms, and they have a total mass equal to that of several million Suns. Whether they are being expelled from the core by a violent event or are the result of some gravitational effect is still unclear.

Over the next decade, astronomers continued to gather clues about the galaxy's heart, sketching a composite portrait first at radio and then at infrared wavelengths. For nearly a century and a half, the astronomical community had regarded infrared radiation as a mere curiosity, but with the advent in the 1940s of better infrared detectors, astronomers at last began to perceive its value. Like radio waves, infrared radiation passes relatively unimpeded through dust clouds; however, since different physical processes generate the two forms of energy, they can yield complementary types of information.

In the late 1960s, astronomers homed in on the galactic nucleus—by then known as Sagittarius A *(pages 69-77)*—with a full arsenal of radio and infrared instruments. At the center, they found a source of extremely energetic radio emission. Unlike the narrow lines produced around the twenty-one-centimeter wavelength by atoms of neutral hydrogen, the signals from Sagittarius A formed a continuous spectrum of radio "noise" that showed no spectral lines. Such emissions, known as synchrotron radiation after manmade particle accelerators, are produced by electrons spiraling around magnetic field lines at close to the speed of light.

Energetic as this radio source seemed to be, it paled by comparison to what investigators found when they turned to Sagittarius A with infrared detectors. In 1968 a pair of researchers working at the Mount Wilson and Palomar observatories discovered that the source at Sagittarius A—which matched its radio counterpart in both location and size—was a thousand times brighter in the infrared than at radio wavelengths.

The logical next steps were to try to understand the nature of this powerhouse and to pinpoint its location. Both were addressed at once. In 1971 astrophysicists at Cambridge University suggested that the dynamo at the core of Sagittarius A was a supermassive black hole. A black hole, by definition, is so compact and massive that nothing—not even light—can escape its gravitational pull, thus rendering it invisible at all wavelengths. Theoretically, however, its existence could be inferred from the effects the black hole would have on its surroundings. The Cambridge team predicted that the intense radiation emanating from gases in violent motion around the hole would be detected as an extremely compact radio source. Three years later, American radio astronomers found just that: a structure of swirling gas with a diameter of only about six light-days.

THE AWESOME VIEW FROM EARTH

With the approach of the twenty-first century, the portrait of the Milky Way grows increasingly detailed as astronomers continue to seek out and penetrate the many veils that hide the galaxy. Enigmas may remain at the distant galactic core, but ingenious theories—coupled with intuitive leaps and careful observations—have resolved many of the mysteries that occupied such investigators as Jan Oort at the outset of his long career. Yet Oort, for one, has never lost the sense of awe that first drew him to study the Earth's galactic home. One night in 1952, his quest for answers took him to an African hillside, where a group of astronomers were investigating possible sites for a new Southern Hemisphere observatory. At some point the other scientists realized—to their consternation—that Oort had disappeared. Fearing that he had been attacked by a wild animal, they mounted an intensive search. After a while they discovered the Dutch astronomer lying flat on his back in the wet grass, staring straight up into the glittering depths of the constellation Sagittarius. Oort shooed away his would-be rescuers. He did not wish to be disturbed. A quarter of a century earlier, he had begun to discern the structure of the Milky Way and had confirmed the way to its heart. Now he was rapt with the simple pleasure of seeing it with his own eyes.

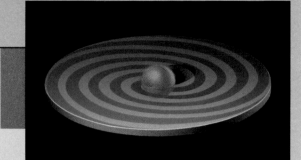

AT THE MILKY WAY'S HEART

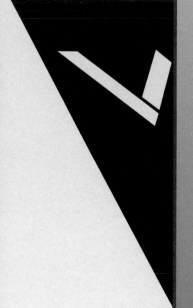

When Harlow Shapley published his studies of globular clusters in the Milky Way in 1918, he concluded that they orbited the center of the galaxy, a region he estimated to be some 65,000 light-years from Earth, toward the constellation Sagittarius. Nine years later, scrutiny of stellar motion by Dutch astronomer Jan Oort confirmed the general location of the galaxy's heart, but its precise distance continued in dispute. There, for many years, the matter rested. Obscured by light-blocking interstellar dust, the galactic core remained an enigma to astronomers using optical instruments.

Then, in the 1950s, came radio astronomy, and suddenly the galaxy's innermost secrets began to reveal themselves. Further advances in observational techniques opened windows on the entire range of the electromagnetic spectrum, from long-wavelength radio signals at one end to the very short waves of X rays and gamma rays at the other. With the help of computers, astronomers can convert this nonoptical information into images for graphic representations like those on the following pages. These images describe a strange arena of swirling clouds and spinning disks, energized by intense magnetic forces, exploding stars, and an invisible engine of unimaginable power— perhaps a black hole.

Behind the Veil,
a Spinning Disk

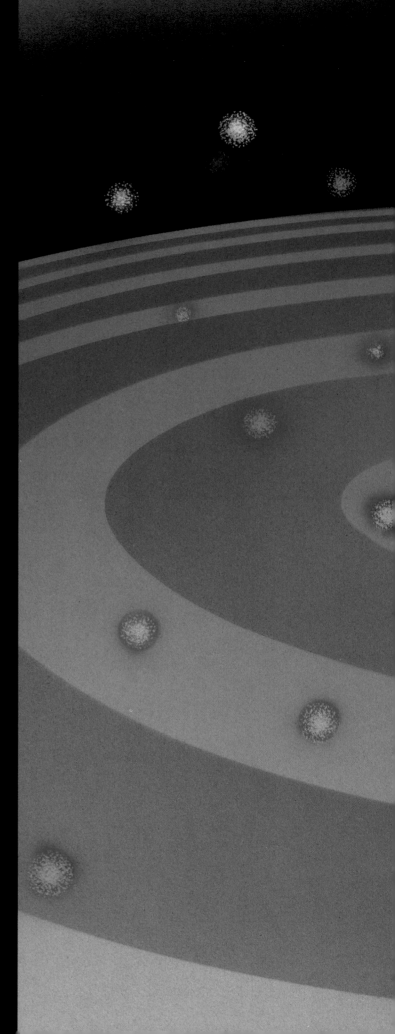

Largely through analysis of radio emissions given off by various forms of hydrogen gas, astronomers in the mid-1970s began to uncover a number of unsuspected features in the Milky Way. One of the first surprises was that the disk defined by the galaxy's pattern of spiral arms was not uniformly flat. As shown here from a perspective above the galactic plane, the Milky Way harbors at its center a secondary disk *(yellow)*, which is tilted at an angle of approximately twenty degrees to the plane.

The inner disk spans 8,000 light-years—a little less than one-tenth the diameter of the main galactic disk *(blue)*—and it nestles well within the nuclear bulge *(lavender)*. Why it should tilt is not yet known, but its movements clearly distinguish it as a unique feature of the galaxy. The study of radio emissions at wavelengths around twenty-one centimeters indicates that the disk is rotating at eighty-one miles per second—about three times faster than the stars and gases in the spiral arms around it—and is expanding outward at ninety-nine miles per second.

This twenty-one-centimeter emission is the product of so-called atomic hydrogen, or HI, one form of interstellar hydrogen. Also called neutral hydrogen because each atom is made up of one positively charged proton surrounded and balanced by one negative electron, HI exists at very low temperatures, around 100 degrees Kelvin. (Abbreviated "K," the Kelvin scale starts at absolute zero, which is −273 degrees Celsius, or −459 degrees Fahrenheit.)

At temperatures above 10,000 degrees Kelvin, atoms of HI are ionized—stripped of their electrons and thus no longer neutral. Known as HII, ionized hydrogen makes up only a small fraction of one percent of the gas in the tilted disk, but, as shown on the following pages, it forms some of the most intriguing structures in the inner galaxy.

The most common form of hydrogen in the tilted disk is molecular hydrogen, or H_2 (because the molecule consists of two atoms bound together). It is found in very cold, dense clouds at temperatures between 10 and 20 degrees Kelvin.

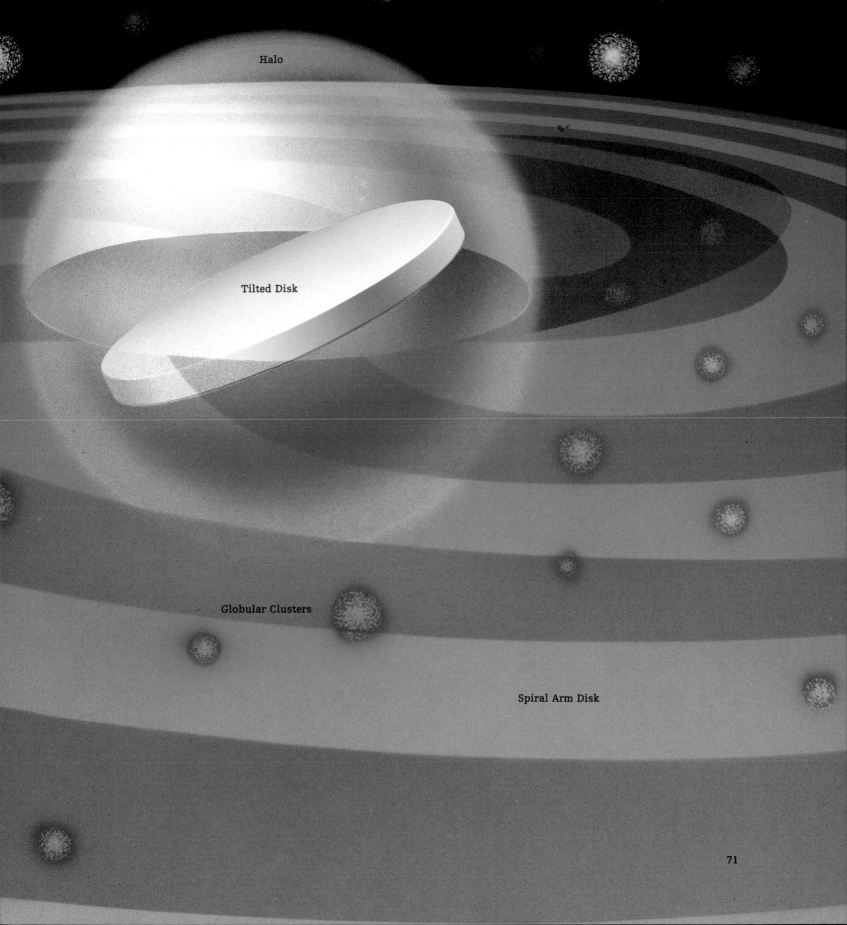

Five areas of thermal *(red)* and non-thermal *(pink and gray)* radio activity within the Milky Way's tilted disk are designated, left to right, Sgr (for Sagittarius) D, B, A, C, and E. Near Sgr B, astronomers have also detected a unique cloud of matter containing molecules not only of hydrogen, which is common throughout the galaxy, but of more than fifty other substances as well, among them carbon monoxide and formaldehyde.

Sgr B

Sgr D

Sgr A

Sgr C

A Cacophony
of Radio Noise

A closer look at the tilted disk reveals a complex structure still largely mysterious in its workings. Shown here and on the following pages are five distinct areas. Collectively known as Sagittarius because their emissions come from the direction of that constellation, they are lettered A through E in order of their discovery. As evidenced by their radio signals, most of them are combinations of different kinds of celestial material and activity.

Dense regions of HII, or ionized hydrogen, produce signals of a type known as thermal. These regions have been heated by radiation from hot stars within them, resulting in a soup of free-floating protons and electrons moving at relatively slow speeds. Near-collisions between the particles release energy in the form of light, heat, and radio waves. Unlike the radio emissions produced by atomic hydrogen, which fall in a narrow range around twenty-one centimeters, those from HII regions cover a broad range of wavelengths. In the radio part of the spectrum, the intensity of these emissions tends to decrease as wavelengths get longer—a unique signature that identifies thermal sources.

An opposite signature—intensity increasing with wavelength—points to nonthermal sources, explained on the next page.

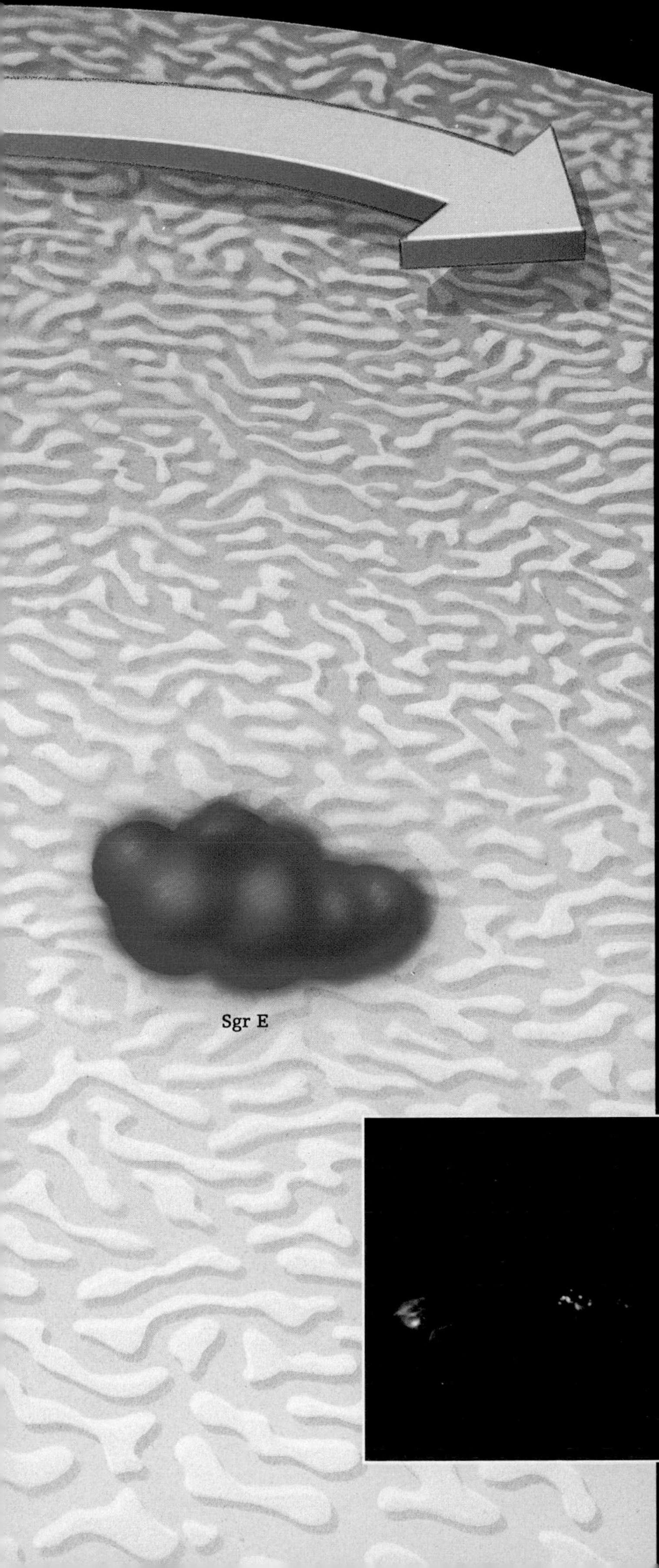

Sgr E

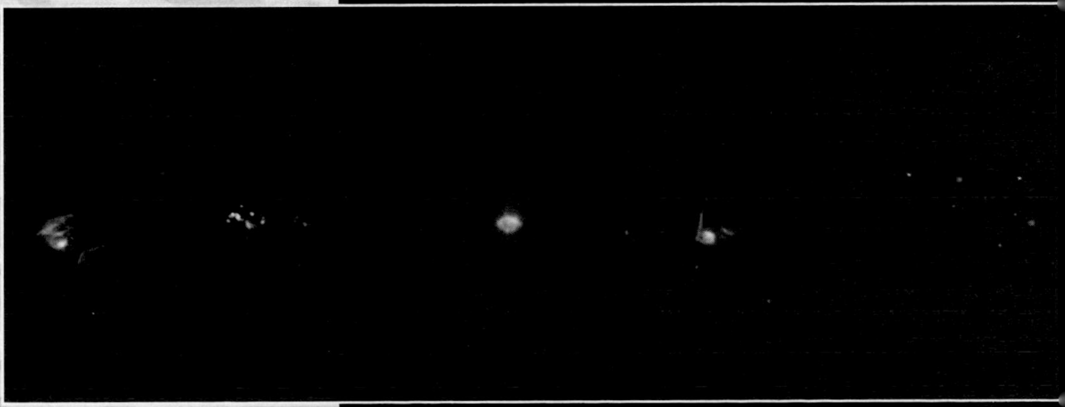

A radiograph of the central region of the Milky Way encompasses the five Sagittarius sources that are depicted graphically at left. Both Sgr A, at the center, and Sgr C, to its right, display filaments projecting from the galactic plane.

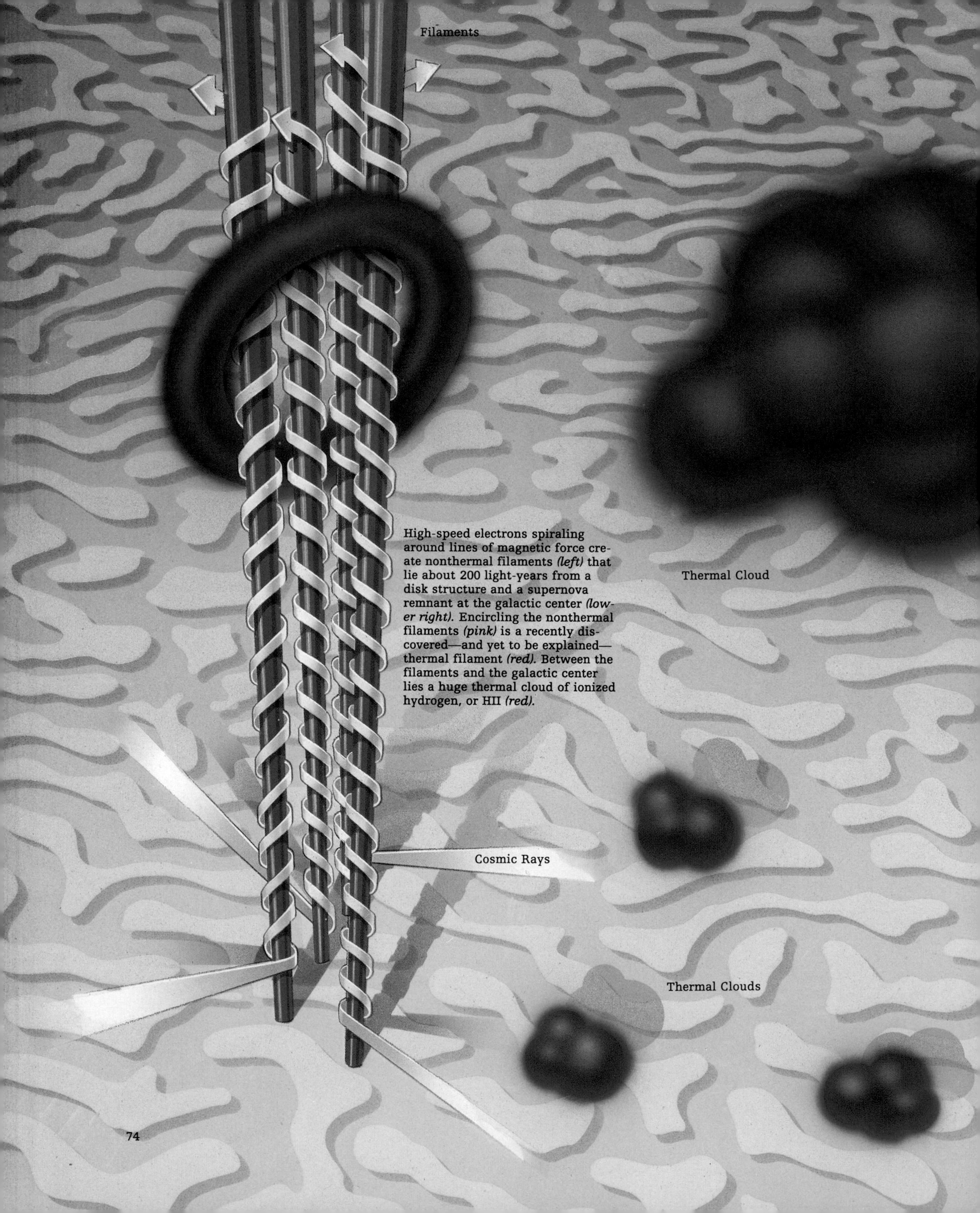

Filaments

Thermal Cloud

Cosmic Rays

Thermal Clouds

High-speed electrons spiraling around lines of magnetic force create nonthermal filaments *(left)* that lie about 200 light-years from a disk structure and a supernova remnant at the galactic center *(lower right)*. Encircling the nonthermal filaments *(pink)* is a recently discovered—and yet to be explained—thermal filament *(red)*. Between the filaments and the galactic center lies a huge thermal cloud of ionized hydrogen, or HII *(red)*.

74

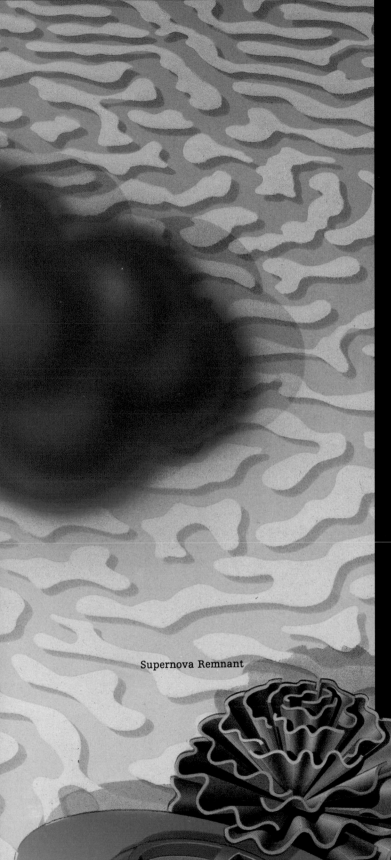

Supernova Remnant

Disk

ZEROING IN ON SAGITTARIUS A

Sharp-eyed radio telescopes are able to discern details in the sky equivalent in size to a dime viewed from twenty miles away. Observed with such an instrument, Sgr A, the first of the Sagittarius sources to be discovered, turns out to be a caldron of thermal and nonthermal activity. Its most striking features are lumpy clouds of heated matter that appear to be associated with nonthermal structures unique to this part of the galaxy: long, thin filaments that grow, like a stand of cosmic trees, at right angles to the galactic plane.

One hundred thirty light-years in length but just three light-years wide, the filaments seem to be shaped by the lines of force of a powerful magnetic field, around which electrons spiral at velocities approaching the speed of light. The spiraling electrons generate a broad range of electromagnetic signals, called synchrotron radiation after a man-made particle accelerator that produces similar emissions.

The origin of the magnetic force and the source of the spiraling electrons are a puzzle. To compound the enigma, astronomers in 1987 detected what seemed to be a thermal filament that inexplicably encircled the tops of the nonthermal strands. Scientists also continue to be intrigued by the very center of the galaxy—shown in greater detail on the following pages—where a supernova remnant has been observed in conjunction with yet another disklike object.

A Black Hole
in a Doughnut

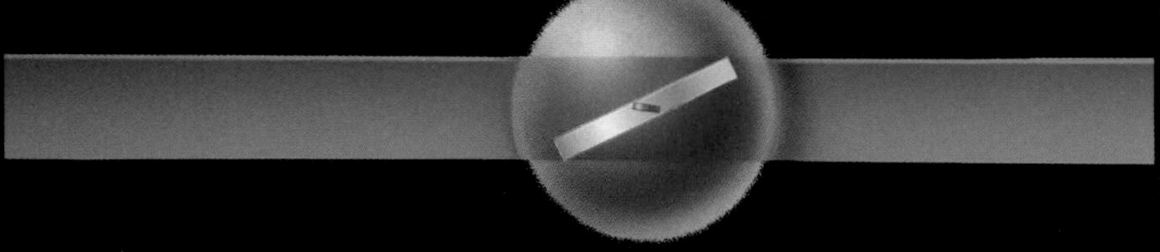

When radio astronomers first glimpsed the central precincts of the Milky Way *(right),* they were amazed at the enormous amounts of energy produced there. In time these signals proved similar to those coming from the centers of several other galaxies; in each case the proposed energy source was a supermassive black hole—an object only about the size of a large star yet containing the mass of at least four million Suns. Its gravity pulls in the surrounding matter, heating it to soaring temperatures and generating radiation all across the electromagnetic spectrum.

But a greater surprise was the discovery in 1984 that the presumed black hole was encircled by yet another disklike object. As shown in the diagram above, this innermost disk nests inside the larger tilted disk *(pages 70-71)* at a forty-five-degree angle.

Unlike its more cohesive larger cousin, the innermost disk is composed of three approximately concentric rings (stylized at right for clarity) that vary somewhat in composition. The outer ring *(orange)* and the middle ring *(dark orange)* are both cool regions, reaching temperatures of only 100 degrees Kelvin. Both are made up of dust, stars, atomic hydrogen (HI), and molecules of hydrogen (H_2) and carbon monoxide, but the middle ring also shows high levels of hydrogen cyanide. Nearest the black hole itself, the inner ring *(red),* composed of ionized hydrogen (HII), ranges in temperature from 10,000 to 20,000 degrees Kelvin.

This complex structure may have arisen when a colony of massive, short-lived stars exhausted their nuclear fuels and exploded. As the surrounding gases were blown outward, a region relatively devoid of matter was left behind—a hole in a doughnut. The stars might then have collapsed into small black holes, and finally into a single monster at the galaxy's heart.

Black Hole

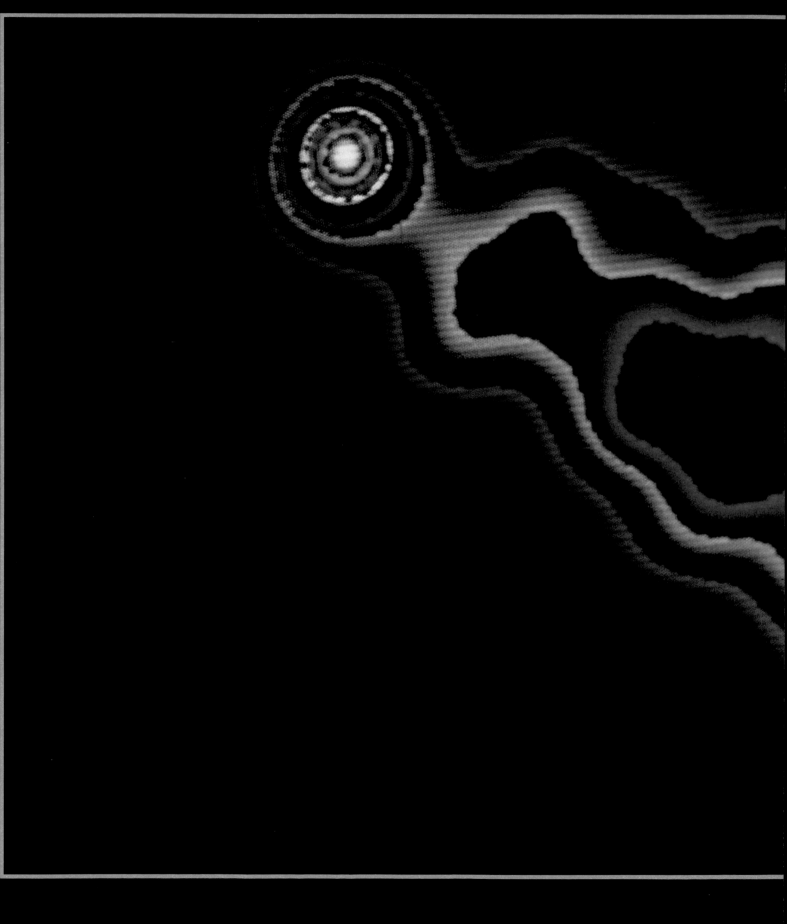

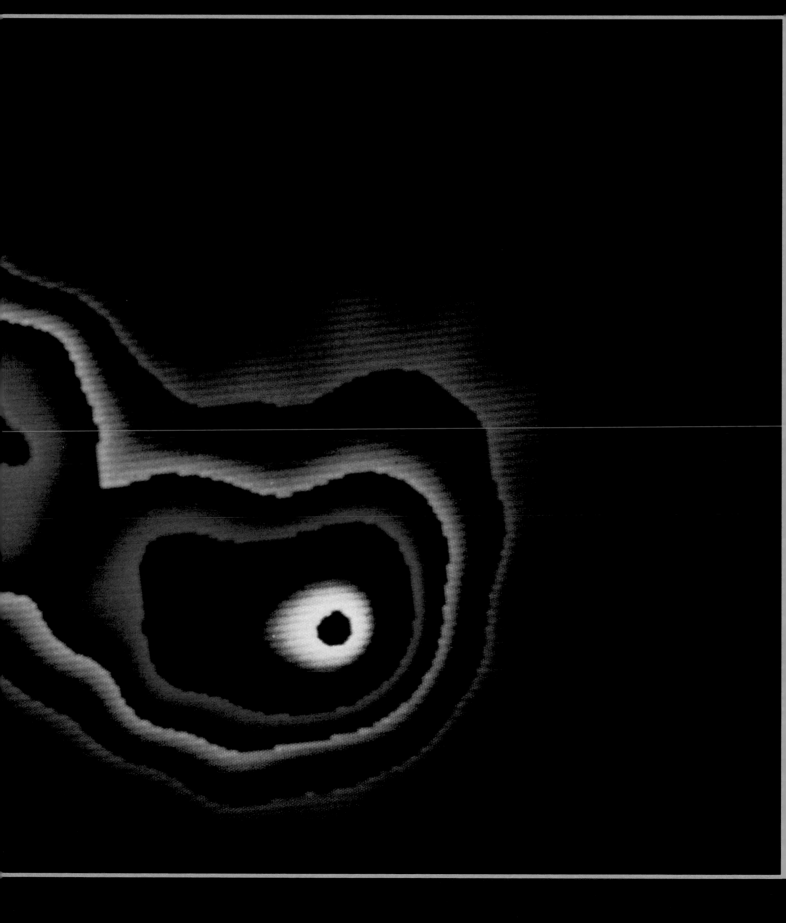

Seven billion light-years from Earth, a quasar designated 3C147 *(lower right)* spews out a stream of matter that is sizzling with energy at its tip *(white, upper left)*. This computer-colored map is based on radio emissions.

ven as astronomers were piecing together the nearer puzzle of the Milky Way in the postwar years, their increasingly powerful telescopes were beginning to gather information about stars and galaxies at unprecedented distances. What these telescopes glimpsed was disconcerting. At the boundaries of the observable universe, it seemed, were objects very different from the celestial population known before then.

A critical link in this process of discovery was forged when an overseas envelope landed on Walter Baade's desk in Pasadena, California, in the fall of 1951. The eminent astronomer, a staff member of the allied Mount Wilson and Palomar observatories, was deeply involved in mapping spiral galaxies in those years, but the message was not one he could ignore. A few weeks earlier, Cambridge University astronomer Francis Graham Smith had used new radio detection methods to locate several strong radio sources in the sky. One particularly clear source lay in the direction of the constellation Cygnus. Graham Smith promptly sent the celestial coordinates to Baade and asked whether he might search for "possible related visual bodies." Without a visual match for the radio signal, and subsequent analysis of the object's light, Graham Smith could not tell if his source in Cygnus was a star or a galaxy, nor did he have any way of knowing whether it was near or far.

Baade immediately set to work with his colleague Rudolph Minkowski—like Baade, a German-born astronomer who had moved to the United States in the 1930s. Minkowski was interested in the universe's more violent phenomena, such as supernovae, and the intense and mysterious radio sources reported by Graham Smith fascinated him. Using Palomar's powerful new 200-inch telescope, the two men began to photograph the tiny areas of sky identified by the British radio instruments.

It soon became clear that only one dim blur in Cygnus matched the coordinates of Graham Smith's radio source, which was later designated Cygnus A. As the first photographs came in, Baade and Minkowski realized that they were looking at something unusual. Although the strength of the radio signal had suggested that the source was probably nearby and possibly a star, the fuzzy, peanut-shaped object did not fit that picture. They then turned to the spectrograph and watched as the fragile glimmer of starlight was spread across a thin strip of film.

Hunched at the eyepiece of Mount Wilson's 100-inch telescope, astronomer Rudolph Minkowski searches for the universe's most energetic objects. Minkowski's interest in galactic oddities led him and colleague Walter Baade to identify the distant radio galaxy Cygnus A in the early 1950s, pointing the way to a whole new class of celestial objects.

The readings were astonishing. Judging by its red shift—the degree to which lines in its spectrum had shifted toward the red end of the visible spectrum—the source lay about a billion light-years away and was receding from the Earth at a speed of about 10,000 miles per second. Its radiation had started earthward when the first algae were growing in terrestrial seas. Yet this remote object was the second-strongest radio source in the sky, after the nearby supernova remnant Cassiopeia A. So much power from so far away meant that Cygnus A must be radiating with the vigor of a hundred billion Suns. The two astronomers concluded that the thermonuclear fusion furnaces churning away inside conventional stars could not possibly produce such an outpouring of energy. A galactic collision was the only explanation they could conceive of.

They were only partially right. Cygnus A may well be a pair of merging galaxies, but that interaction does not of itself produce the object's intense radio energy. At the heart of Cygnus A is an unknown, unseen, yet immensely powerful engine.

Baade and Minkowski had discovered the first of what would later be called active galaxies. The celestial bodies collectively known by that name form a gallery of cosmic violence. Many have fantastically bright nuclei. Some emit prodigious jets of matter and radiation from their cores. The most dramatic, now known as quasars, may be as small as our own Solar System yet blaze forth with the light of a hundred Milky Ways.

So confounding are these celestial dynamos that for many years, disputes raged about their distance from the Earth. One astronomical camp, doubting that sources so energetic could be both small and very remote, challenged the accuracy of the first optical identifications and argued that the objects must reside within—or certainly near—the Milky Way. The opposite camp, while unable to explain how anything could produce such powerful radio and light emissions, accepted what the red-shift numbers said: that the sources were scattered to the very edge of the universe.

One thing was clear in all the furor. Whatever the path to an answer, it was most likely going to involve the emergent science that had started all the trouble in the first place—radio astronomy.

THE OTHER UNIVERSE

Although Karl Jansky and Grote Reber had pioneered radio astronomy in the United States during the 1930s and early 1940s, the new observational technique was pursued most vigorously after the war by two independent teams in England and another in Australia.

Cambridge University hosted a group at its Cavendish Laboratory. There, Francis Graham Smith and his colleagues were led by Martin Ryle, a spare, boyish-looking Englishman who later won the Nobel Prize for his accom-

Built in 1947, this rudimentary radio telescope at England's Jodrell Bank observatory was a spider web centered on a 126-foot antenna. Wires strung from perimeter posts to the antenna's base formed the framework of the bowl (here outlined in red); more wire around the frame made up the reflecting surface of the skeletal dish. Astronomers pointed the antenna by pulling on the guy wires that held it upright.

plishments. About a hundred miles away, near the Cheshire village of Jodrell Bank, a second British team, directed by Bernard (later Sir Bernard) Lovell, established the University of Manchester's Nuffield Radio Astronomy Laboratories. Lovell, a physicist whose devotion to building telescopes was rivaled only by his passion for cricket, would turn Jodrell Bank into a renowned center for radio astronomy. In 1947 he and his group completed the world's first great radio antenna, a 218-foot-wide dish intended at first to detect radio emission from particles known as cosmic rays—fast-moving atomic nuclei raining earthward from the Sun and unknown sources in space.

The Australian group was a government research effort headquartered in Sydney and led by Joseph Pawsey, whose early career had been in commercial radio and television broadcasting. In 1946 he began to study solar radio emissions and sunspot activity.

Most of the British researchers had also begun their careers in the practical applications of radio transmission. Unlike the stereotype of astronomers as pallid night creatures glued to the eyepieces of their telescopes, these were sturdy young men who favored sweaters, tweed jackets, and baggy trousers. Trained for almost everything except astronomy, they had been drawn into the wartime development of radar, Britain's early warning system against German bombers. (The name *radar* is an acronym derived from "radio detection and ranging.")

When they turned onto the path of pure science, their first tools were familiar—aerials donated as war surplus from the radar establishment. "I remember most the slogging out to our observatory four times every twenty-four hours through most of a year to reset those Wurzburg antennas and the chart recorder," wrote Francis Graham Smith of his early days at Cambridge University's radio-telescope facility. "Above all, perhaps, days of honest toil in good company. It was a long time before we even realized that we were astronomers."

Before the 1940s the known radio emanations of the universe consisted of a few snarls and whistles spread more or less randomly across the sky. "Cosmic static," some people called it. During the war the bursts of radio energy proved troublesome, sometimes disrupting Allied radar transmission. An enterprising British army engineer named James Stanley Hey traced the problem to sunspots. When the war ended, he stayed with the army and began to assemble a map of the radio sky.

Hey and his colleagues in the Army Operational Research Group soon found that the Sun was by no means the only discrete source of celestial radio energy. He noted that at certain places in the heavens, radio transmission intensified abruptly, evidence that a powerful signal was originating there. But Hey was stymied when he tried to relate his finds to optical features in the Milky Way. His equipment was incapable of pinpointing the source of an emission so that he could link it to any one star.

Radio astronomers share with their optical colleagues the problem of res-

THE WHIRLPOOL'S MANY FACES

Astronomers study galaxies, such as the aptly named Whirlpool *(right)*, by examining them not only in the familiar light of the visible spectrum but also at shorter and longer wavelengths—X ray, ultraviolet, infrared, radio. As shown here, each type of radiation reveals different features.

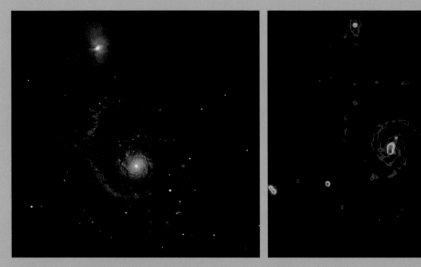

Visible. At the wavelengths of ordinary light, the Whirlpool is a glowing spiral of young, bluish white stars, with one of its arms tugged slightly askew by the gravitational pull of a smaller neighbor.

Radio. Long-wavelength radio waves often indicate regions of nonthermal radiation caused by electrons spiraling around compressed magnetic fields. Radio intensity is greatest at the galaxy's center.

olution—of receiving signals clearly enough to distinguish one star from another. The route to better resolution is straightforward: The bigger a telescope is, relative to the wavelength of the radiation it captures, the greater its ability to discern detail. But this is a harsh equation for radio astronomers. While wavelengths of light are reckoned in millionths of an inch, the radio spectrum begins with wavelengths of about one inch and stretches out to waves that are miles long. A good optical telescope can bring into focus a small coin roughly two and a half miles away. The same kind of resolution in a radio telescope would require a dish 150 miles wide.

USEFUL INTERFERENCE

Because building such enormous telescopes is utterly impractical, the groups at Cambridge, Jodrell Bank, and Sydney all opted for the next best thing: an interferometer, so called because such an instrument causes radio waves to mix and interfere with each other in useful, measurable ways. An interferometer could be devised by the simple tactic of connecting two dishes with electrical cables or radio links. Signals from radio waves rippling across each dish would travel to a central processor, where they would interfere with each other in such a way that matching signals could be combined to produce a sharp, clear image. In its resolving power (but not its sensitivity), the interferometer would behave like a single enormous antenna as wide as the distance from one component dish to the next.

Inspired by Martin Ryle—a man so intensely involved with his studies that one colleague said Ryle "knew what it felt like to be an electron in the Sun"—

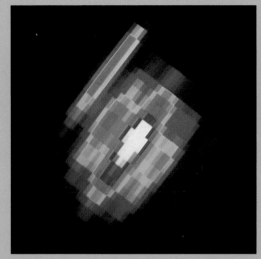

Infrared. At shorter wavelengths, interstellar dust clouds that are warmed by nearby stars become visible in varying concentrations.

Ultraviolet. Usually linked to highly energetic new stars, ultraviolet radiation peaks *(red)* at the galaxy's core and along two of its arms.

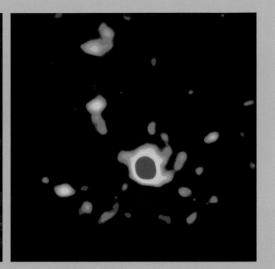

X ray. Intense, very short wavelength radiation from the galaxy's center may come from gas clouds in a region of active star formation. Isolated galactic hotspots *(yellow)* may be binary systems made up of a star orbiting—and feeding—a black hole.

the Cambridge team worked feverishly to cobble together an interferometer from their leftover wartime equipment. In 1945 they created a crude instrument consisting of two small aerials that were carried by hand about the university rugby ground. By 1948 they had built larger antennas in misty fields a few miles from the university, despite being hampered by thorny hedges and horses that trampled the cables into the dirt.

In Australia, Pawsey's group met the interferometry challenge with an ingenious design that, in effect, used the Tasman Sea as one of the antennas: Radio waves reflecting off the flat surface of the sea were picked up by an aerial mounted on a cliff top near Sydney. That aerial also received waves directly from space, and the two sets of signals interfered with one another, with the result that they created a sharp image. Meanwhile, the Jodrell Bank group relied on their huge radio dish until they could find a suitable location for a second antenna.

The race to carry out the first real radio reconnaissance of the universe was on. The three teams scanned the skies, aware that their colleagues were doing the same, looking for specific radio sources and hoping that their telescopes could resolve the sources well enough to allow the radio emissions to be matched to known visual objects. Soon the Australian group had a solid result. They narrowed down an area in Cygnus that Hey had found, and they identified three more sources: one, Taurus A, within the Milky Way's own Crab nebula, and the other two, Virgo A and Centaurus A, in other galaxies. Meanwhile, Ryle and Graham Smith at Cambridge had turned their paired antennas away from the Sun and were compiling their own impressive list. In 1950 the Cambridge catalog of the Northern Hemisphere held fifty sources of radio emission.

As the teams collected radio sources, they all realized that they were looking at two different varieties. One kind of signal came from extended, diffuse celestial objects—probably whole galaxies—while a second type appeared to stem from much smaller, pointlike sources. The enigmatic little powerhouses were called radio stars.

Arguments erupted over just what and where the supposed radio stars were. In 1951 and 1952, the team at Jodrell Bank noted that the distribution of radio stars in the sky was similar to the distribution of known galaxies, suggesting that radio stars fit best into an extragalactic pattern. This of course raised the question of whether they were stars at all. But Ryle and his Cambridge team strongly believed the discrete radio sources to be within the Milky Way, arguing that no one had convincingly matched the sources with optical galaxies. The Cambridge group gained a reputation for insularity, and Ryle himself was later described as both tight-lipped and chary of new information. Wrote Robert Hanbury Brown, a Jodrell Bank staff member, ''It was part of the conventional wisdom at Jodrell Bank that Cambridge had only three standard reactions to our work: (1) 'it is wrong,' (2) 'we have done it before,' or (3) 'it is irrelevant.' ''

But the tide inexorably began to turn in favor of the extragalactic nature of radio stars. Cambridge's own Francis Graham Smith may have dealt the killing blow to the argument for their being located in the Milky Way when he supplied Baade and Minkowski at Palomar with the information that placed Cygnus A at such an enormous distance. Then, in 1955, the Cambridge team of astronomers finally capitulated: Martin Ryle acknowledged in a paper that at least some radio stars existed at extragalactic distances.

Still, as late as 1958, the number of positive optical identifications of radio sources was pitifully small. Of the more than 2,000 radio sources cataloged by the Cambridge and Sydney teams, only forty-six were linked to visual objects. Most of them had been identified as either normal galaxies, very hot stars, or supernova remnants. Seven were extragalactic mysteries—like Cygnus A, astonishingly small and lively. These remote points of light, sizzling with radio energy, would soon cause astronomers and the rest of the world to marvel at the strangeness of the universe.

ANDROMEDA BY RADIO

Located about two million light-years from the Milky Way, the Andromeda galaxy—portrayed optically at left—is considered a galactic neighbor. It was thus one of the first celestial objects to be studied when radio astronomy was born in the 1940s.

THE FIRST QUASARS

By 1956 Hanbury Brown and his colleagues at Jodrell Bank had transformed the lab's 218-foot dish into one-half of an interferometer of unusual length, linking the dish to an antenna located twelve miles away, next to England's most elevated pub, the Cat and Fiddle in Derbyshire. But despite their long interferometer, the Jodrell Bank team still could not pinpoint the celestial coordinates of the farthest radio sources. To do that required considerable legwork by one of their most enterprising members.

Astronomer Henry Palmer took to the road. Motoring across England with several small, steerable antennas, he would stop whenever a promising open space presented itself. Down country lanes and across muddy fields (and once, Palmer recalled later, "in a private paddock, where a nice daughter provided tea and buns"), he would turn his antennas toward the same object targeted by the big dish and radio the signals back to Jodrell Bank for analysis. "In my mind's eye," wrote Hanbury Brown of his intrepid colleague, "I always see Henry Palmer in wellington boots."

Palmer's efforts brought impressive results. By 1961 he had created the interferometric illusion of a radio telescope more than seventy miles across and able to resolve sources equivalent in angular size to a mile-wide object at the distance of the Moon. The Jodrell Bank astronomers could now take a closer look at Cygnus A.

The object delivered another grand surprise: Instead of radiating energy uniformly into space, Cygnus A sent out its radio signals from two vast branches, or lobes, extending from a visible galaxy that was also a radio

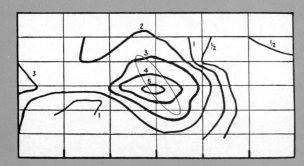

Radio map, 1950. Hand-drawn by Jodrell Bank astronomers, a contour map of the Andromeda galaxy shows the best resolution available from a first-generation radio telescope in 1950. Superimposed in red is the outline of the visible galaxy.

Radio map, 1985. Produced by the twelve-telescope Westerbork array in the Netherlands, this map of Andromeda illustrates a vast improvement in the resolving power of radio telescopes. Contour lines join areas of equal intensity; the darkest regions are the most intense. (Patterns on either end of the spiral are unrelated to the galaxy.)

source. The symmetrical radio lobes sprawled across a distance in space of approximately 320,000 light-years, rendering Cygnus A one of the largest single entities in the known universe. Consisting of electrons and other sub-atomic particles spiraling within a hot, diffuse gas, the lobes seemed to have been expelled by something that was positioned in the center of the galaxy, something that could accelerate the particles nearly to the speed of light and send them coursing trillions of miles through the weak interstellar magnetic field, emitting radiation that caused deep space to crackle at radio frequencies. The structure baffled astronomers. What could generate such acceler-ating force within a galaxy's core?

The answer did not come for years, until scientists were able to collect and categorize enough odd specimens in their celestial bestiary. In the late 1950s even Palmer's new and accurate measurements were not good enough to allow an optical identification of the smallest radio objects. Then in 1960, Jodrell Bank researcher Cyril Hazard—a slender, laconic thirty-four-year-old—pro-posed an ingenious new technique that would improve resolution many times over. By tracking the Moon continuously as it passed in front of, or occulted,

Tireless watchers of the radio sky, the instruments above represent only a portion of the so-called Very Large Array, which was built in 1980 in the desert near Socorro, New Mexico. Stretched out in a Y pattern across seventeen miles of desert, the twenty-seven dishes of the VLA combine to surpass the resolving power of the best optical telescopes.

a radio source, and by measuring the precise moments when the radio signals vanished and reappeared, the position of the source upon the sky could be pinpointed. Hazard tested the method successfully that December—and took it with him when he moved to Australia to join the Sydney team.

There the young astronomer would soon have a chance to use his technique. In 1961 the Australian team completed construction of a two-million-dollar, eighteen-story radio telescope, a behemoth whose 210-foot dish shadowed the wheat fields of Parkes, New South Wales. The telescope seemed ideal for Hazard's lunar occultation method, and a choice opportunity was near at hand. In April, August, and October of 1962, the Moon would pass across the fixed position of 3C273, an intense radio source (so named because it was object 273 in the *Third Cambridge Catalogue* of radio sources). But a test run of the telescope revealed a major problem. The source would be eclipsed directly over the dish and reappear sixty-two degrees farther down. The dish, unfortunately, could only drop sixty-one degrees before hitting the ground.

Hazard and John Bolton, director of the telescope, came up with a wonderfully simple solution: Just dig a trench to accommodate the rim. As for the safety mechanisms that stopped the dish from dropping more than sixty-one degrees—simply disconnect them. And the cogs that kept the machine from hitting the ground? Tear them off. And teach some workers to hand crank the telescope in case the power failed. In short order, the newly built miracle of high technology was a jury-rigged device dedicated to one task only: deciphering 3C273.

When the moment of truth arrived, the lunar occultation method worked perfectly. The signals from 3C273, boosted through an amplifier, were recorded in long inked scribbles on tapes at the telescope's recording station. Hazard gave one set of records to a colleague and kept another, and the two men flew separately to a computer center 200 miles distant. Even if one plane crashed, reasoned Hazard, the tapes would survive.

Australia's team in Sydney was able to translate the inked tapes into celestial coordinates and tentatively matched them with a very faint starlike object in the constellation Virgo. But the final stage in the quest to identify this radio source was left to the Americans and their 200-inch mirror at Palomar Observatory.

At this time a younger generation of researchers was moving into the California observatory in the footsteps of Baade and Minkowski. Among them were Thomas Matthews, Allan Sandage, and a Dutch astronomer named Maarten Schmidt. Hazard sent his coordinates to Schmidt, who passed them on to Matthews, who matched them to a starlike object on a photographic plate that had been made by Sandage.

MISSING PIECES

The California scientists were baffled by 3C273. The starlike object could not be identified as either a galaxy or a star. It had an inexplicable bright spoke extending from it. Most puzzling of all, its spectrum, which would tell the

THE DISCOVERY OF QUASARS

During 1962 British astronomer Cyril Hazard *(right)* started the chain reaction of discoveries that led to the first known quasar, named 3C273. By pinpointing the coordinates of a mysterious radio source, he enabled Thomas Matthews at California's Palomar Observatory to find an optical match: a dim stellar object with a peculiar jet *(center)*.

astronomers how far away the object was, how fast it was moving, and something about its chemical composition, looked like so much gibberish. They could not find the characteristic pattern of absorption and emission lines that served as the key to every spectrogram.

This was not the first such puzzle Allan Sandage had seen. Two years previously he had come across a similar radio source known as 3C48. The dim point of light had looked like a star, with an odd luminous cloud—a "nebulosity"—on two sides of it. Although Sandage would come to be known as something of a visionary, a mystic with a grand view of cosmology reminiscent of his mentor, Edwin Hubble, he could make no sense of 3C48's light. "It was the weirdest spectrum I'd ever seen," he commented later. The strange object radiated most strongly in the far blue and ultraviolet fringes of the visible spectrum. Stranger still, it had shown only the slightest trace of hydrogen, the primary ingredient of normal stars.

Photographs revealed another peculiarity as well. The brightness of 3C48 varied from night to night. Because the speed of light is taken to be a universal speed limit, an object cannot vary in brightness faster than light can travel across it. If 3C48 could blaze and fade overnight, it had to be small enough for light to cross it in a day—that is, it could be no more than about twice the diameter of the Solar System. If this object existed at extragalactic distances, it would have to radiate many times more powerfully than the brightest galaxies known. The puzzle of 3C48 was merely compounded when 3C273 joined it in the category of unreadable sources.

AN EYE FOR RIDDLES

The answer to the spectral riddle came to Maarten Schmidt, Sandage's young colleague, in a flash of insight. The lanky, bespectacled Dutchman had only recently begun studying "peculiar" stars, but he had been in love with as-

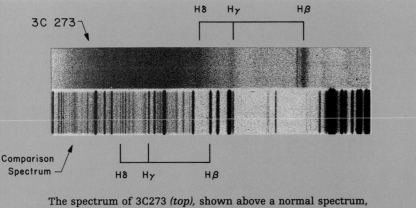

3C 273

Hδ Hγ Hβ

Comparison
Spectrum

Hδ Hγ Hβ

The spectrum of 3C273 *(top)*, shown above a normal spectrum, was illegible until Maarten Schmidt in 1963 recognized that its four hydrogen lines were drastically red-shifted—one of them invisible in infrared.

tronomy since building his own telescope from a toilet paper roll at the age of twelve. His fellow astronomers knew him as a "good eye," someone who could really work a telescope. On February 5, 1963, that eye saw something no one else had seen. As Schmidt sat at his desk and studied the tiny scrap of film bearing the spectrum of 3C273, the gibberish was suddenly intelligible: Schmidt had spotted the distinctive spacing of four hydrogen emission lines. The lines had been shifted so far toward the red as to be almost unrecognizable, like familiar faces seen in improbable surroundings.

Finding the hydrogen lines showed the way to other emission lines of ionized magnesium and oxygen. As Schmidt stared at the slip of film, colleague Jesse Greenstein walked past his office door. "Jesse," Schmidt called out, "I think there's a 16 percent red shift in 3C273." This meant that the object was receding at nearly 16 percent of the speed of light, placing it almost three billion light-years from Earth. Given its distance, the small body also had to be about a hundred times brighter than any other luminous radio source that had so far been identified.

Once Schmidt had taken the intuitive leap on 3C273, Jesse Greenstein and Tom Matthews figured out the red shift of their other puzzle, 3C48, in about fifteen minutes. Where 3C273's red shift had moved the starlike object well clear of the Milky Way, the shift for 3C48 placed it at the very edge of the known universe. Its spectral displacement was 37 percent, representing a recession speed of almost one-third the speed of light and a distance of some five to six billion light-years.

ANNOUNCING SOMETHING AWFUL

Together, 3C48 and 3C273 appeared to announce the existence of a new celestial species. The odd galaxy Cygnus A had been remarkable enough, but at least it had looked like a galaxy in its photographs. These objects, though, seemed too far, too small, and too powerful to be anything that astronomers had ever known before. That night, Schmidt told his wife, "Something awful happened at the office today!" When she looked alarmed, the Dutchman realized that he had chosen the wrong English word. "I mean," he amended, "something wonderful."

The strange objects would come to be called quasars, a name derived from "quasistellar radio source," and the term was accepted by the astronomical community with a kind of professional shudder. By the mid-1960s, it was clear that the enigmas were not very stellar, nor were they, for the most part, radio sources. As astronomers surveyed other suspicious stellar objects, looking for extreme red shifts, they collected more and more quasars. By studying them with optical, radio, and X-ray telescopes, they were able to identify a distinct set of characteristics: Quasars are a hundred to a thousand times brighter than normal galaxies; they have large red shifts, implying both great recession speed and vast distance; they are compact and blue; they emit X rays; and their spectra show very bright emission lines, meaning that the gas in quasars is very hot. Only about 10 percent are strong radio sources. (The first ones discovered were all radio emitters simply because astronomers were only looking for radio sources in those days.) Some quasars resemble Cygnus A in having radio or optical jets—narrow beams of radiation-emitting particles—ejected from their cores, and many vary rapidly in brightness.

SHINING HEARTS OF VIOLENCE

Most astronomers today recognize quasars as the most energetic members of a whole group of peculiar celestial objects with bright nuclei. The first were discovered as early as 1943, when Carl Seyfert, a postdoctoral fellow at Mount Wilson Observatory, reported strong, distinctive emission lines in some galactic spectra. These "Seyferts," as the galaxies were called, are all disk galaxies—a generic term that includes both flattened, well-developed spirals like the Milky Way and similar disks without prominent spirals—and all have brilliant, starlike nuclei and spectra that point to some violent activities in their hearts. As other windows opened in the electromagnetic spectrum, some Seyferts turned out to be powerful radio and X-ray emitters, while others shone most intensely in the infrared.

No less exotic than Seyferts are galaxies known as double radio sources. Cygnus A, the peculiar sight that so surprised Walter Baade and Rudolph Minkowski in 1951, is one of these. Through an optical telescope, they look like elliptical galaxies, but radio maps reveal that two lobes of radio-emitting matter are spraying out from their centers. The lobes can be of stunning size; for example, in one galaxy, 3C236, they span almost 18 million light-years, dwarfing the Milky Way as a whale does a minnow. The energy radiating from the radio lobes of a typical double radio source equals that released by 10 billion supernova blasts.

Like Seyferts, but slightly dimmer, are N-galaxies, so called because they have bright nuclei. N-galaxies are often highly variable, sometimes brightening and dimming in a matter of months. This category includes an even brighter and more wildly fluctuating subspecies known as BL Lacertae objects, named for the prototype found in the "lizard" constellation, Lacerta. When an observer managed to block out the light from the original BL Lacertae's bright center and take a reading on the fuzzy glow around it, the spectrum resembled that of a typical elliptical galaxy. Its red shift put it at about one billion light years from Earth. Several dozen counterparts, known collectively as BL Lacs, have been found since then.

Maarten Schmidt's analysis of the red shift of 3C273 yielded a distance of almost three billion light-years. Subsequent discoveries placed quasars at distances ranging from about one billion to 15 billion light-years.

The astronomers who collected and named these galaxies did not know what could cause the galactic centers to blaze, erupt, and oscillate the way they do. And quasars were the biggest mystery of all. Unlike Seyferts or BL Lacs, they could not even be seen in any detail. The most promising theory suggested that quasars were the brilliant hearts of galaxies so far away that their dimmer galactic regions could not be perceived against their blazing nuclei. Indeed, Maarten Schmidt's first paper in 1963 described 3C273 as the nucleus of a distant galaxy. However, he could muster no persuasive optical proof.

THE RED-SHIFT CONTROVERSY

"That period was incredibly electric," Allan Sandage has said, recalling the years from 1960 to 1966. "Every time you went to the telescope and came down you had a major new discovery."

Within two years of Schmidt's breaking the spectroscopic code of 3C273, astronomers were announcing quasar red shifts of 200 and 300 percent, indicating distances up to 12 billion light-years and recession speeds exceeding 90 percent of the speed of light. Physicists realized that if quasars were as far away as these red shifts seemed to show, yet as small as their variability indicated, then they produced more power than anything else in human view. The enigmatic objects asked a lot from astronomers. For some, they asked too much.

Rather than accept the powerhouses at face value, a few astronomers preferred to believe that the red-shift measurements did not apply to quasars—that the objects were in fact much closer than their red shifts appeared to indicate. This suggestion struck at a fundamental belief of modern astronomy, namely, that red shift is a result of the expansion of the universe—that it is, as astronomers say, cosmological, indicating both velocity and distance. If red shift did not work for quasars, the thing had some undiscovered flaw, and the whole of astronomy must quake until the flaw was found.

One of the first alternative notions to be offered in the early 1960s was that the quasar red shift was not a Doppler shift caused by the expansion of the universe but a product of the gravitational drag on photons escaping from the great mass presumed to be in the quasar. Schmidt and Greenstein demolished the idea in a paper published in 1964. They presented two convincing arguments. First, neutron stars—stupendously dense collapsed stars in which a cubic inch of matter would weigh 100 million tons if brought to Earth—generate gravitational red shifts of less than 10 percent, much less than those measured for the two quasars. Second, the emission lines observed in quasars could not be produced in that kind of gravitational environment.

Other theorists ventured that quasars owed their high red shifts to their being spewed out of the Milky Way and nearby galaxies at tremendous speeds. Although it was attractive, this model foundered on one critical issue: No one had ever chanced to observe any blue-shifted quasars, the

Jet

Jet

Accretion Disk

Black Hole

Magnetic Field Line

At the hypothetical core of a radio galaxy, a spinning black hole drags gaseous debris into a flattened, revolving accretion disk. Matter in the disk falls toward the center, generating a jumble of magnetic field lines *(red arrows)* that align themselves along the black hole's spin axis. This axis defines north-south escape routes for the electromagnetic energy and high-speed particles, forming two opposed jets.

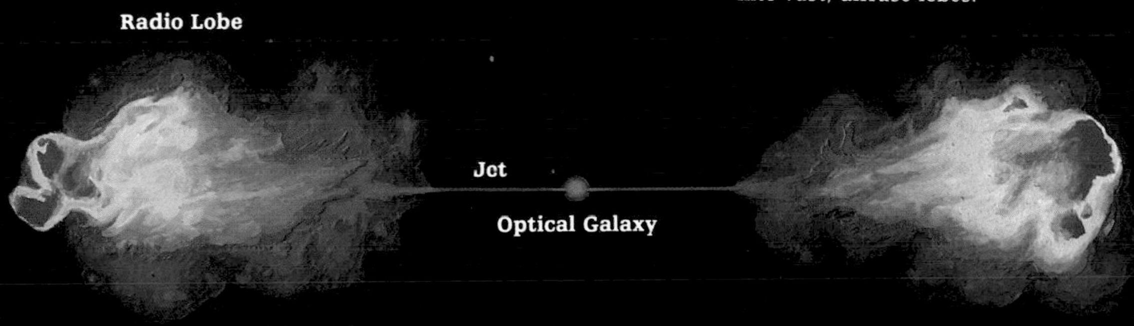

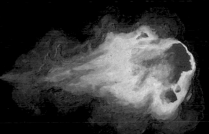

Dwarfed by its sprawling lobes, the optical component of a radio galaxy sends out jets at near light-speed from the Solar System-size engine *(opposite page)* that lies buried in its core. Several million light-years from the core, the jets slow abruptly as they encounter ionized intergalactic hydrogen, producing shock waves and energetic "hot spots" *(red)* that spread the narrow jets into vast, diffuse lobes.

Radio Lobe

Jet

Optical Galaxy

A CELESTIAL DYNAMO

Among the most intriguing objects in the universe are radio galaxies, distinguished by a pair of enormous, diffuse lobes that pour forth radio waves. The lobes are often linked to a central, compact radio source by concentrated jets of matter and energy. The whole complex stretches as much as 18 million light-years from one side to the other. (Viewed at optical wavelengths, radio galaxies are visible only as an ordinary-looking star system that spans about a hundred thousand light-years.)

The engine at the heart of a radio galaxy has never been directly observed, but the likeliest candidate is illustrated at left: a rotating, supermassive black hole, fed by stray gas drawn from supernovae, stellar exhaust, and passing galaxies. The black hole's tremendous gravitational force—equivalent to that of hundreds of millions of Suns—would draw these gases into a so-called accretion disk, where convection, friction, and electromagnetic turmoil can generate electric potentials of a hundred million trillion volts near the system's center. Channeled in opposite directions along the black hole's axis of rotation, this energy could then whip outward at nearly the speed of light to form the galaxy's twin-lobe signature *(above)*.

ejecta of other galaxies hurled in our direction.

Almost from the start, the prickly mantle of devil's advocate went to Halton Christian Arp, a young astronomer from Mount Wilson and Palomar observatories. Arp was a decidedly unconventional soul who lived in a Moorish stone house built by a magician. Where others looked for the rules underlying celestial phenomena, Arp looked for the exceptions. During the early 1960s, he began to compile his photographic *Atlas of Peculiar Galaxies,* portraits of strange celestial objects that had caught his eye. It was while studying these phenomena that he developed the idea that quasar red shifts were not caused entirely by their velocities of recession.

One of Arp's chief arguments against a cosmological red shift for quasars lay in photographs showing apparent physical links between objects of greatly different red shifts. Some quasars appeared to be bound to nearby galaxies by a glowing trail of gas; some galaxies seemed to be deformed by the influence of a passing quasar. The most compelling case, that of the galaxy NGC 4319 and the seemingly nearby quasar named Markarian 205, appeared to show a quasar with a seven-percent red shift (which is quite low for a quaar) connected by a bridge of luminous material to a galaxy with a red shift more than ten times lower.

Astronomers have since virtually ruled out the possibility that the bridge is a real physical structure, concluding that it is an optical illusion caused by a slight overlap of the two objects' halos in the photograph. Many also believe that the universe holds so many millions of galactic combinations that these visual coincidences are inevitable. Arp continues to battle for the noncosmological red shift, however—a thorny reminder to his colleagues that the matter of quasars is far from completely settled.

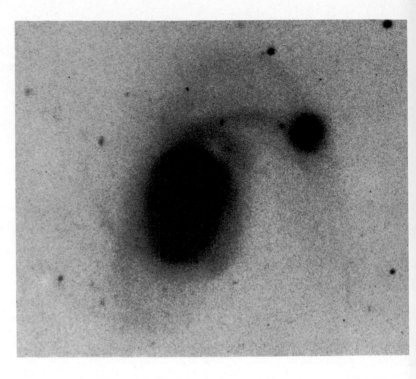

Ambiguous evidence of an anomaly, this photograph appears to show a physical link between NGC 7603—a type of active galaxy known as a Seyfert—and a smaller companion. Red-shift studies of both objects indicate that the smaller object is receding from the Earth at twice the speed of the larger, leading gadfly astronomer Halton Arp to claim that red shift is an unreliable indicator of distance. His critics maintain that Arp's galaxies are joined only by photographic coincidence.

FINDING FUZZY QUASARS

Nevertheless, further support for the conventional, cosmological theory of red shift has come with improved optical telescopes. In the mid-1970s, optical astronomers began to use a new light-detecting mechanism called the charge-coupled device, or CCD, on large ground-based telescopes. The small silicon chip is fifty to a hundred times more sensitive to light than the best photographic emulsion. Using the CCD, optical astronomers looking at a variety of quasars could just begin to discern a surrounding fuzz of luminosity—a fuzz that might be billions of stars. By the early 1980s, observers had identified about two dozen of these so-called fuzzy quasars, objects embedded in a nebulous substance that had the appear-

THE ACTIVE GALAXY ROSTER

Although active galaxies have many names, they may represent stages in the evolution of a single kind of galaxy with a black hole at its center. Common types are listed below.

Quasars are the smallest, brightest, and generally most distant of the active galaxies. They emit high levels of radiation, often as visible and infrared light and X rays, and sometimes as radio waves.

N-galaxies are elliptical, with nuclei that are small and blazingly bright.

BL Lacertae objects, a type of N-galaxy, have brilliant nuclei and vary rapidly in luminosity.

Radio galaxies emit huge, fast-moving jets of radio energy from a small nucleus.

Seyferts are disk galaxies; like N-galaxies, they have very luminous hearts. Their spectra indicate violent activity in their centers.

ance, intensity, and color one would expect to see in a galaxy. Moreover, the fuzz had the same red shift as the quasar.

This evidence strongly supports the theory that quasars lie at the center of galaxies. The peculiarities among the various types of active galaxies—the N-galaxies, the BL Lacs, the double radio sources—must still be studied and decoded by astronomers, because the oddities carry crucial messages about the nature, origin, and fate of these objects. But it appears that active galaxies constitute a kind of gradient of youthful, evolving galaxies on which quasars shade gradually into Seyferts in a single, albeit complex, process. As Allan Sandage puts it, "Seyferts and quasars are the same phenomenon, just with the rheostat turned up."

HEARTS OF DARKNESS

If it is true that quasars are the nuclei of young galaxies, astrophysicists must still explain where these nuclei get their power. The thermonuclear reactions within stars, while extremely energetic by human standards, are not up to producing the intense and sharply fluctuating radiation of an active galaxy.

The most commonly accepted theory about quasars places a black hole at their cores. At the center of a galaxy, such an object might work as an inexorable drain, creating a cosmic maelstrom of matter and energy—called an accretion disk—spiraling about the lip of the black hole and accelerating crazily the closer it comes to the vortex. Gas, dust, and even stars that are drawn into this whirlpool would be torn apart by the enormous gravitational forces there and propelled into the center, friction heating them as they circle inward, causing them to give off the radiation that is then seen as the X-ray, radio, and optical emissions of quasars.

A supermassive black hole, concentrating the masses of a billion Suns, would need to consume ten solar masses a year to maintain a quasar's brightness. An engine with such an appetite cannot run forever—or even for very long, by cosmic standards. Astronomers find increasing evidence that when a black hole's fuel is exhausted, the great engine may be replenished by passing galaxies. Although that process would unfold over tens or hundreds of millions of years, and so cannot be observed from start to finish, during 1987 three teams of astronomers, working independently at Palomar and at Arizona's Kitt Peak, found three separate cases in which a passing galaxy seemed to be coming close enough to a quasar to arouse the drowsing monster by feeding it stars. These investigators also believe that such mergers among galaxies might have ignited quasars in the ancient universe, when galaxies were closer together and collisions among them more frequent.

In that more compact universe, some 15 billion years ago, the night sky would have trembled with hot, young stars. Nearby galaxies would have been as easily visible as the Magellanic Clouds are today, and quasars would have been close enough to look like bright stars. If quasars are in fact creatures from another epoch—faint lights reaching us from a time when the universe was young—the visible universe must have an edge, beyond which no more quasars would appear.

Until the late 1980s, some astronomers believed that this cosmic fence—a limit to space and time—had been located. In 1973 a quasar was discovered with a red shift of 350 percent (and a distance therefore exceeding 12 billion light-years), and for more than a decade afterward, no higher red shifts were found. Then, in 1986, astronomer Patrick Osmer at the National Optical Astronomy Observatories in Tucson, Arizona, and colleagues at Cambridge University began using a novel laser-scanning device—the Kibbelwhite Machine, named for its inventor—to isolate high-red-shift quasars on photographic plates. Since then, Osmer and his British partners have reported quasars with red shifts of more than 400 percent. Light from the quasars would have traveled for about 15 billion years, roughly 90 percent of the current estimate for the age of the universe.

The idea of discerning the birth of the universe has irresistible mystical overtones. "The best text that could be indicated here," said Allan Sandage in a 1972 lecture on the red-shift limit, "would be that in the beginning there was darkness upon the deep. There was light, and out of that light came everything that we now observe."

WHEN GALAXIES MEET

ver the past two decades, astronomers have gathered increasing evidence that galaxies are unlikely to pass their lives in splendid isolation. Many show signs of violent interaction with fellow star systems—wrenching near-misses or ruinous collisions. The likelihood of galactic encounters is relatively high: The star systems tend to occur in pairs and groups, where they are separated from one another by only 10 to 100 galactic diameters. Thus, they do not move at random but typically orbit around a shared center of gravity. The pull exerted as two galaxies sweep past each other can resculpt them both. More severe interactions may cause their orbits to shrink and the galaxies themselves to combine.

As illustrated on the following pages, the evidence for galactic meetings is both observational and theoretical. Many of the bizarre-looking galaxies lumped together in the category known as "peculiar" seem to be made up of components that display distinctly different characteristics: parts that spin in opposite directions, for example, or that consist of dissimilar populations of stars. On the theoretical side, scientists have simulated galactic collisions and produced the very forms of turbulence observed in the skies, including bridges of stars between galaxies and long stellar tails streaming off into space.

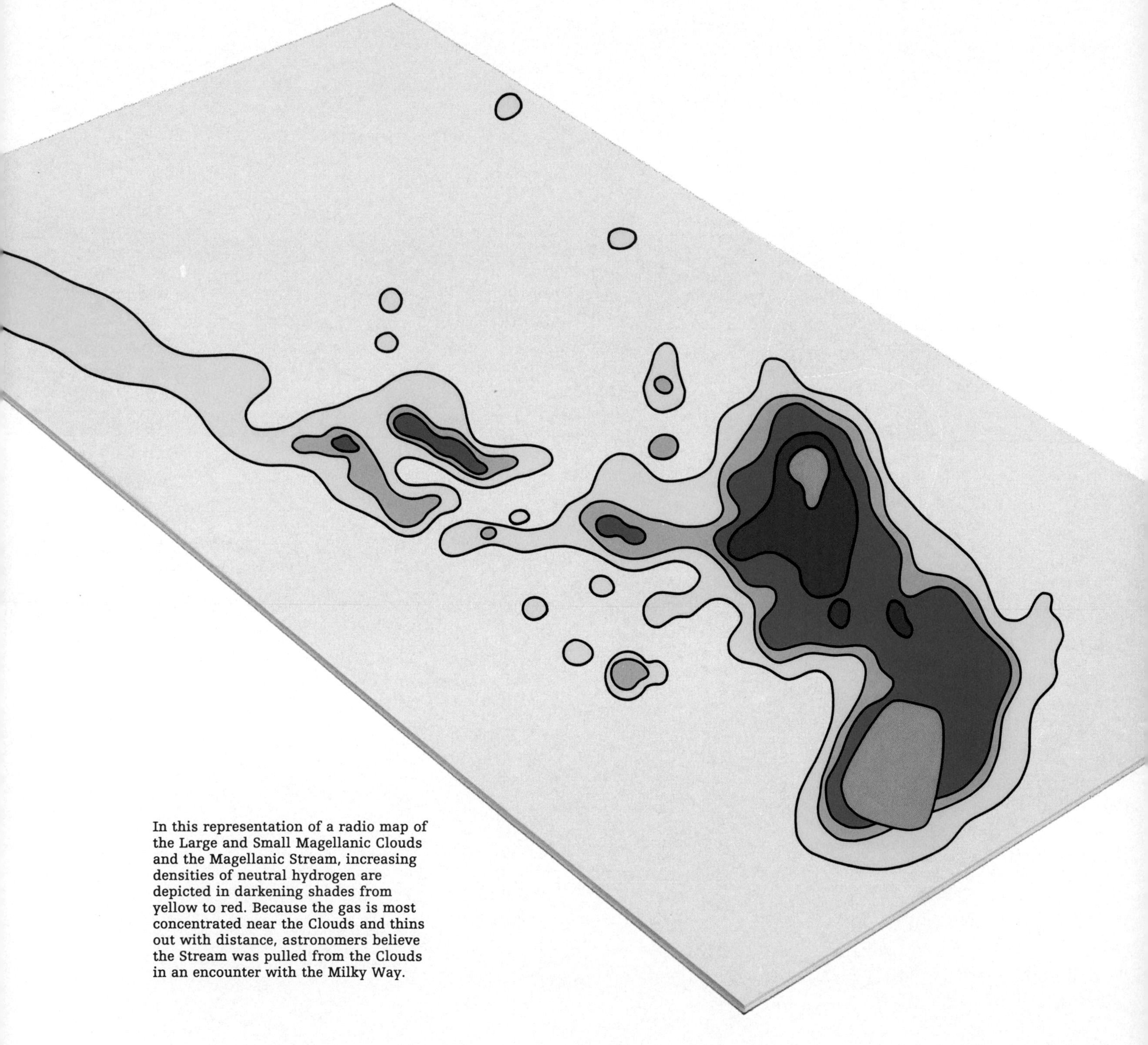

In this representation of a radio map of the Large and Small Magellanic Clouds and the Magellanic Stream, increasing densities of neutral hydrogen are depicted in darkening shades from yellow to red. Because the gas is most concentrated near the Clouds and thins out with distance, astronomers believe the Stream was pulled from the Clouds in an encounter with the Milky Way.

The Sun and its planets are bit players in an intergalactic tug-of-war that has been going on for millions of years. Hints of it appeared in 1973, when radio astronomers found a long, narrow column of neutral hydrogen gas extending about 300,000 light-years from the Large and Small Magellanic Clouds toward the south pole of the Milky Way *(below)*.

This so-called Magellanic Stream strongly indicates that the much larger gravitational force of the Milky Way is deforming the galaxy's two nearest neighbors. Some astronomers also believe that greater violence is in store for the pair of galactic satellites: The smaller systems may ultimately be swallowed by the Milky Way. The Large and Small Magellanic Clouds are also affecting each other. A bridge of hydrogen gas between the diminutive galaxies seems to mark a near-miss some 200 million years ago.

Further evidence of interaction lies in the Milky Way itself. Astronomers have detected a slight warp in the plane of the galaxy's disk—extrusions of neutral hydrogen on opposite sides of the bigger star system that roughly line up with the Clouds' gravitational pull.

Extrusions of neutral hydrogen gas *(purple)* suggest a probable close encounter between the Milky Way and the Large and Small Magellanic Clouds. The Magellanic Stream trails from the Clouds toward the south pole of the Milky Way, while on either side of the Milky Way's disk, a warped lip of hydrogen suggests the gravitational attraction exerted by the Clouds on the larger star system. A hydrogen bridge between the Magellanic Clouds indicates interaction at even closer quarters.

Large Magellanic Cloud

Small Magellanic Cloud

OF TIDES AND TAILS

In the encounter between the Milky Way and its much smaller neighbors *(pages 100-101),* the Milky Way itself was only lightly affected. But when galaxies of nearly the same mass interact, a radical metamorphosis results, producing odd-shaped galaxies such as the ones on the opposite page.

The primary force in a meeting of equals is tidal, analogous to the force the Moon exerts on the Earth. Because lunar gravitational attraction tugs hardest on the near side of the Earth, oceans on that side are drawn moonward, creating a bulge of water; halfway around the planet, a similar bulge occurs because water there is, in effect, left behind as the Moon drags on the Earth itself. In much the same way, the mutual gravitational attraction of roughly equal-size galaxies disturbs the arrangement of stars and gas in each galactic body, stretching both systems out of shape. The closer the two come during their meeting, the stronger the tidal force is. Their final configuration is greatly affected by small variations in the angle or velocity at which the galaxies approach one another and by the direction in which they rotate around their respective galactic centers.

As shown here in illustrations based on a computer simulation, the galaxies pull bridges of stars from each other on their near sides and leave long tails of stars trailing from their far sides. (To simplify the variables in such simulations, scientists reduce complex spiral galaxies to simple disks.)

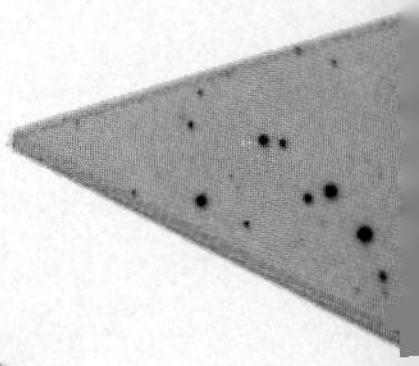

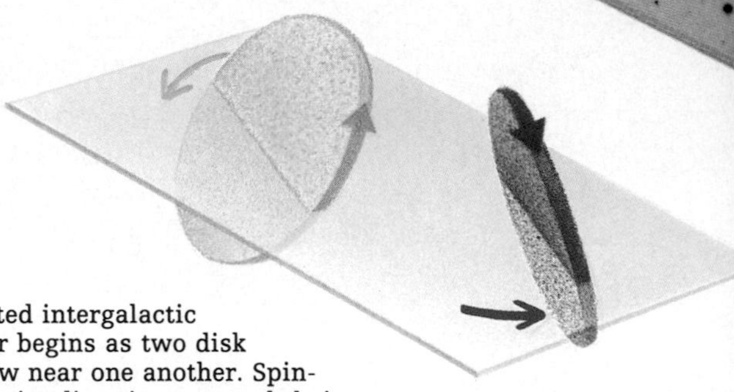

1 A simulated intergalactic encounter begins as two disk galaxies draw near one another. Spinning in opposite directions around their respective galactic centers, they begin to orbit their common center of mass.

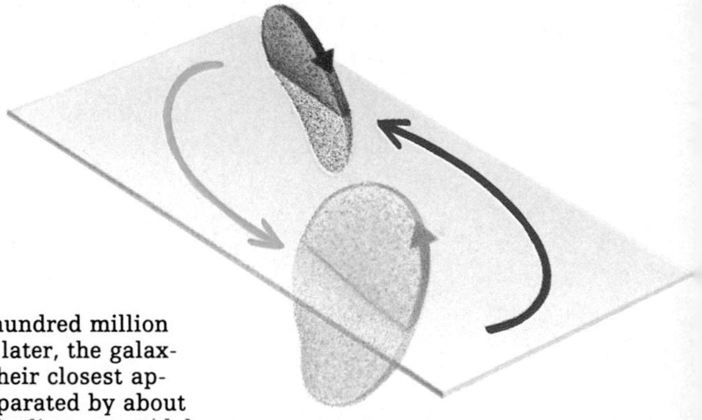

2 Two hundred million years later, the galaxies make their closest approach, separated by about one galactic diameter. Tidal forces have begun to distort the disks on both the near and far sides.

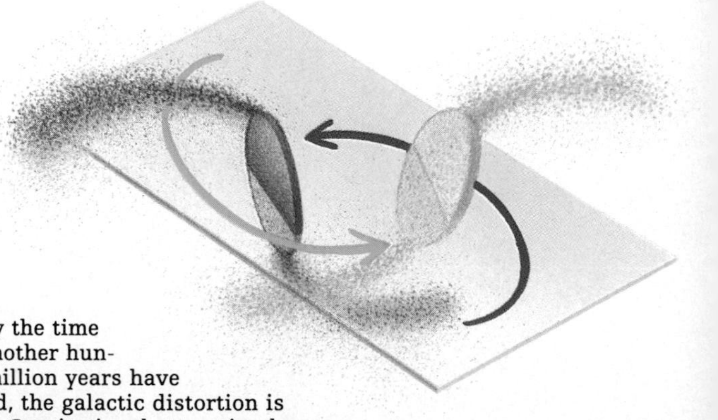

3 By the time another hundred million years have elapsed, the galactic distortion is severe. Gravitational attraction has tugged out a bridge of stars linking the two galaxies together, and stellar streamers have begun to trail behind each of the partners in the dance.

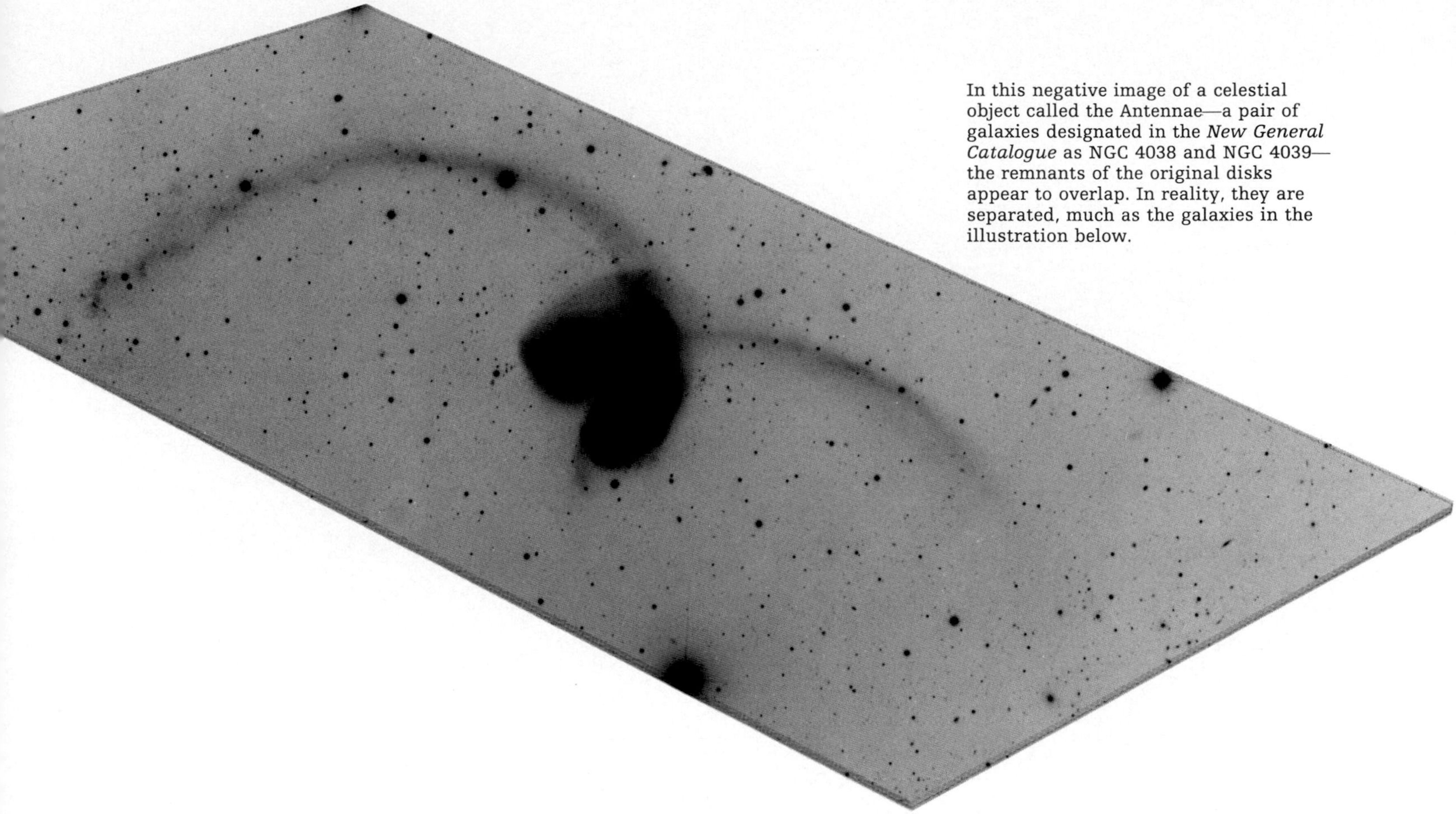

In this negative image of a celestial object called the Antennae—a pair of galaxies designated in the *New General Catalogue* as NGC 4038 and NGC 4039— the remnants of the original disks appear to overlap. In reality, they are separated, much as the galaxies in the illustration below.

4 Two hundred million years later, each partner's starry tail has lengthened to several times the diameter of the parent disks.

5 After the passage of another two hundred million years, the diameters of both disks have diminished even further. Some of the stars in the tails have gained so much velocity in the encounter that they have escaped into intergalactic space.

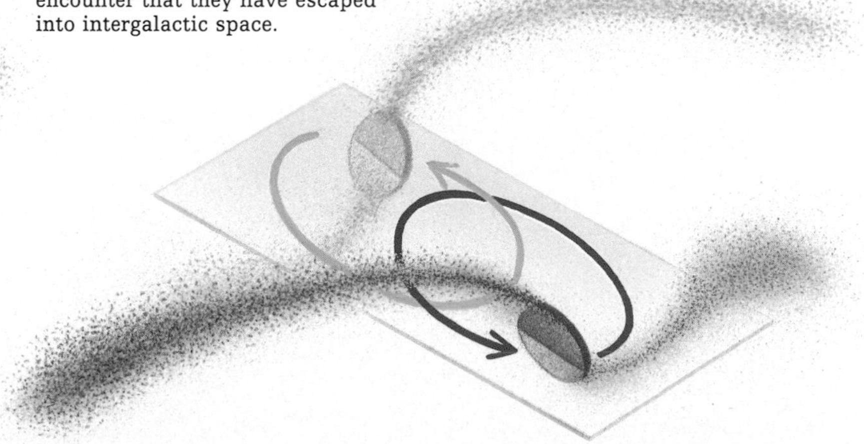

A galaxy designated NGC 7252
is believed to have formed when
two large disk galaxies began to merge
a billion years ago. The central body
resembles a conventional elliptical
galaxy, but the two long tails of stars
are remnants of the two disks.

1 Spinning in opposite directions, two large disk galaxies,
each about the size of the Milky
Way, approach each other head on in
this depiction of a computer simulation.
At this stage in their engagement they
are about two galactic diameters apart.

2 About 600 million years later,
the two galaxies collide. The
probability of stars actually hitting
each other is vanishingly low because
stars are even more widely spaced than
a few dozen baseballs scattered across
the whole of North America.

5 After the second collision, the two
galaxies separate again. As shown
here, their maximum separation for this
round is only a quarter the initial
distance. Stars from both of the galaxies
have begun to mix, but two distinct
entities remain visible.

THE PATH TO A MERGER

As long as equal-size galaxies approach at an angle and stay about one galactic diameter apart, they keep some individual identity. If the parties meet head on, however, the likely result is a complete merger.

As the galaxies pass through one another, stars in one system exert a gravitational attraction on their counterparts in the other, creating a drag force known as dynamical friction, similar to the drag of the Earth's atmosphere on an orbiting satellite. The result is orbital decay, a shrinking of the galaxies' orbits around their common center of mass. Changes in the galactic orbits produce changes in the gravitational fields within the galaxies, affecting the orbits of individual stars. In a process called violent relaxation, the stars in the two galaxies are, in effect, shuffled together.

Astronomers have found evidence of such mergers in the form of faint shells around giant elliptical galaxies. These dim wrappings seem to consist of stars much like the old stars in the disks of spiral galaxies. Coupled with observations of ellipticals with twin tidal tails, the shells lead scientists to think that some—and perhaps all—large elliptical galaxies are the products of galactic collisions. Even more intriguing is the possibility that some mergers may give birth to the highly energetic objects called quasars.

3 Following the initial collision, the two galaxies recede from each other, but because of dynamical friction, they separate by less than one galactic diameter. Mutual gravitational forces distort the plane of each disk into a caplike shape.

4 About a billion years after their first approach, the galaxies collide again, drawn together as their orbits decay. Most of the galaxies' mass now lies at their common border, but some stars, boosted by the encounter, are slipping free of the parent bodies.

6 As the stars mingle, their mutual attraction changes the gravitational field in the merging system. The two original galaxies are still in evidence, but the stars are beginning to move around the more massive center of the new galaxy.

7 Two billion years after their initial approach, the two original disk galaxies have lost their identities and become a single elliptical galaxy.

Evanescent Galactic Rings

Among the most exotic of so-called peculiar galaxies are vast bracelets of stars know as ring galaxies. Astronomers believe that these fragile-looking objects are the product of direct collisions between a disk galaxy and a much smaller intruder. In effect, the intruder scores a galactic bull's-eye, causing stars in the target disk to rush toward a temporary new center of gravity. As the intruder punches through the disk and out the other side, the disk stars halt their inward motion; the now diminishing gravity causes them to rebound, forming a large, expanding ring.

Ring galaxies are rare for two reasons. First, the conditions that create them are unusual: The intruder must approach at a particular angle and strike the disk more or less dead center. Second, once the event has occurred, its effects are relatively transient on the cosmic scale; the stars in a ring will disperse and the ring vanish within a few hundred million years.

1 A small spheroidal galaxy approaches a much larger disk galaxy that is about the size of the Milky Way. Moving along an orbital path nearly perpendicular to the plane of the disk galaxy, the spheroid will strike to the right of the disk's center.

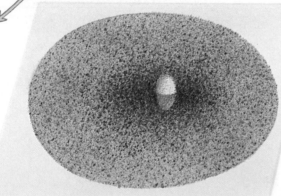

2 As the smaller galaxy hits the disk, its extra gravitational pull causes stars throughout the disk to rush toward the center. For a time, the disk appears to have two bulges—its own plus that of the intruding sphere.

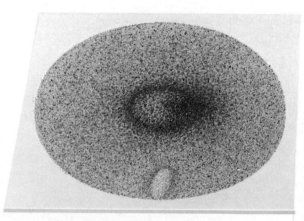

3 Several tens of millions of years later, the intruder galaxy has passed through the disk and is continuing along its orbital path. The stars it attracted toward the disk's center have begun to rebound and are forming a small but expanding ring.

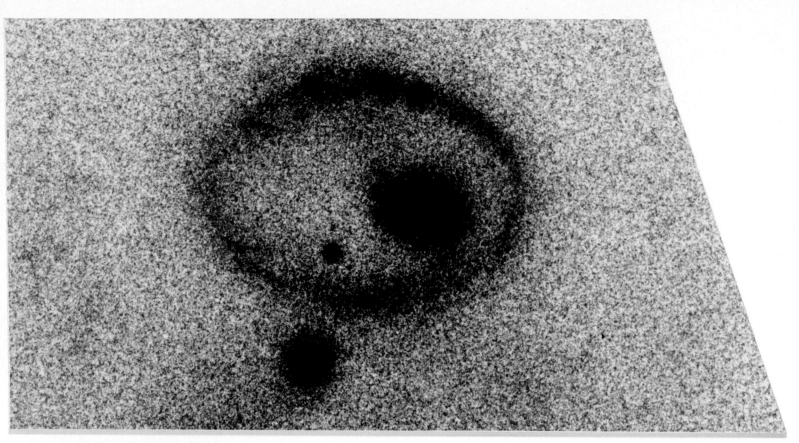

A galaxy designated II Herzog 4 is one of a few dozen known ring galaxies. At lower center is the small galaxy whose passage through a disk galaxy is believed to have created the ring, much as in the simulation shown below.

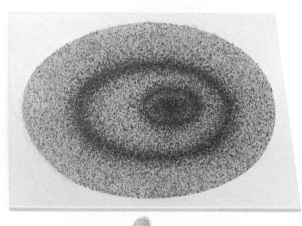

4 As the ring of stars expands, it compresses gas and dust, triggering star formation. Illuminated by new stars, the ring stands out against the dimmer background of the disk.

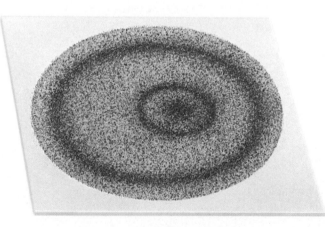

5 Over time, the stars in the ring will begin to disperse, and the ring will eventually vanish altogether, leaving the disk galaxy much as it was before the intruding spheroid made its way through.

Galaxies in Turmoil

Computer models demonstrating how certain galaxy formations might result from different kinds of galactic interaction embrace only a fraction of the varieties of peculiar galaxies to be found in the heavens. Nonetheless, astronomers believe that many of the unclassifiable galactic shapes they observe—including those presented here—are the products of close encounters, collisions, and mergers.

One type that gives clear evidence of its origin is the so-called polar-ring galaxy, a spindle-shaped core encircled by a ring *(below)*. Astronomers have yet to model all the details of the encounter, but they have determined that the apparent spindle is in fact a disk galaxy viewed nearly edge on. That the ring rotates at right angles to the plane of the disk, over its galactic poles, strongly suggests that it is the remnant of a separate star system. The two components may also display other signs of individuality. Some disks show relatively little star formation but are circled by polar rings bursting with new stars.

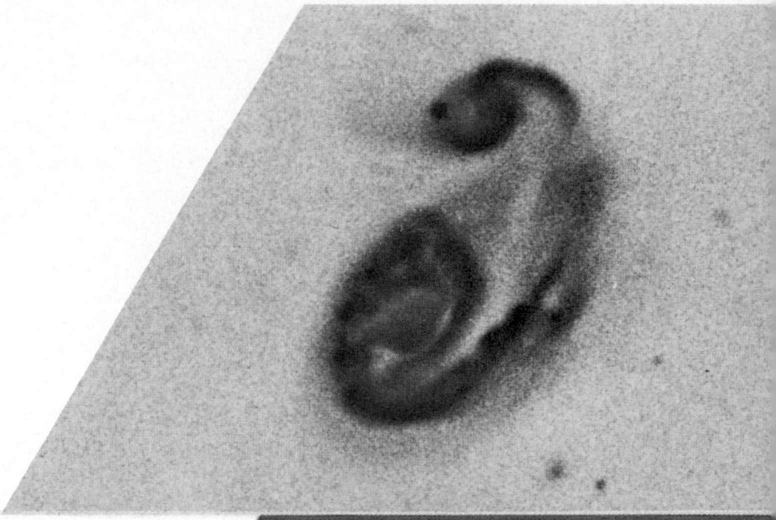

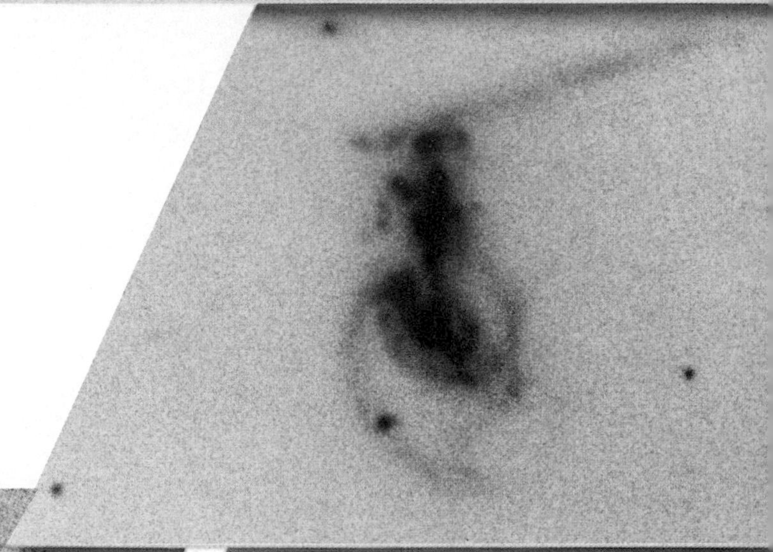

Among the plethora of peculiar galaxies pictured here, polar-ring galaxies *(below)* most clearly reveal evidence of their parent disk galaxies *(diagram)*.

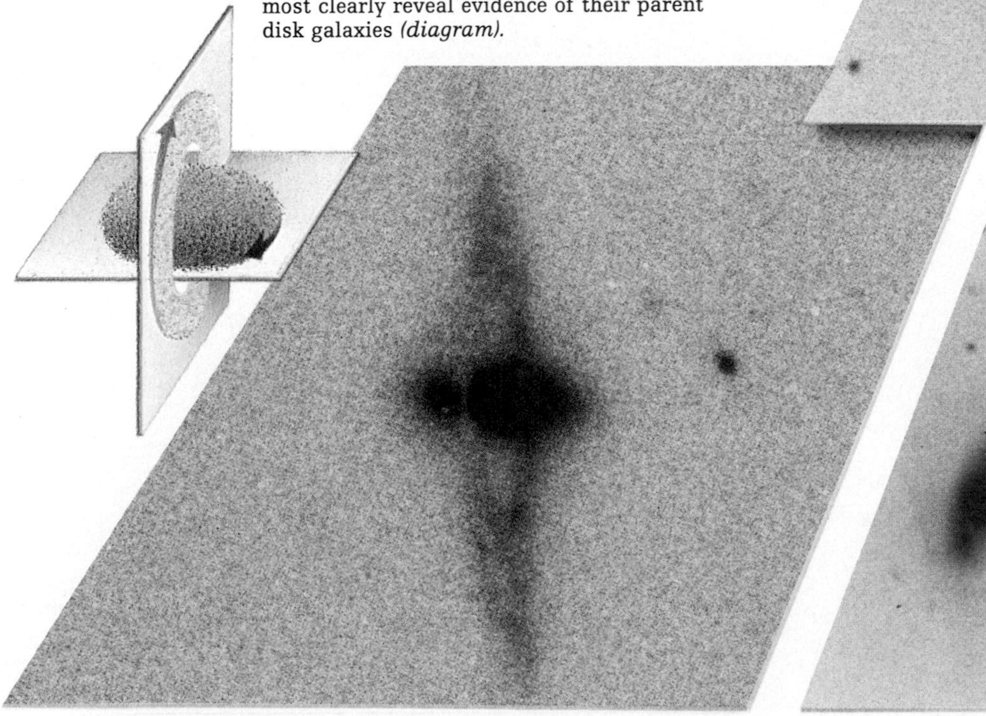

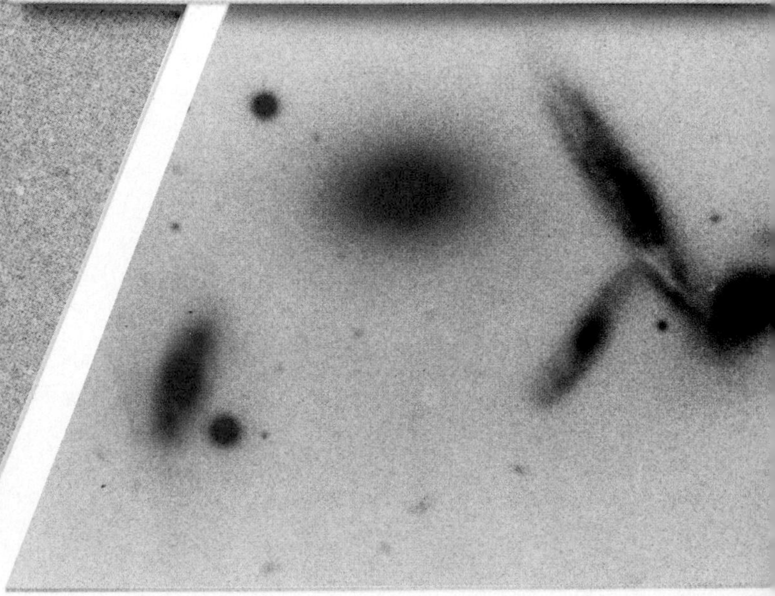

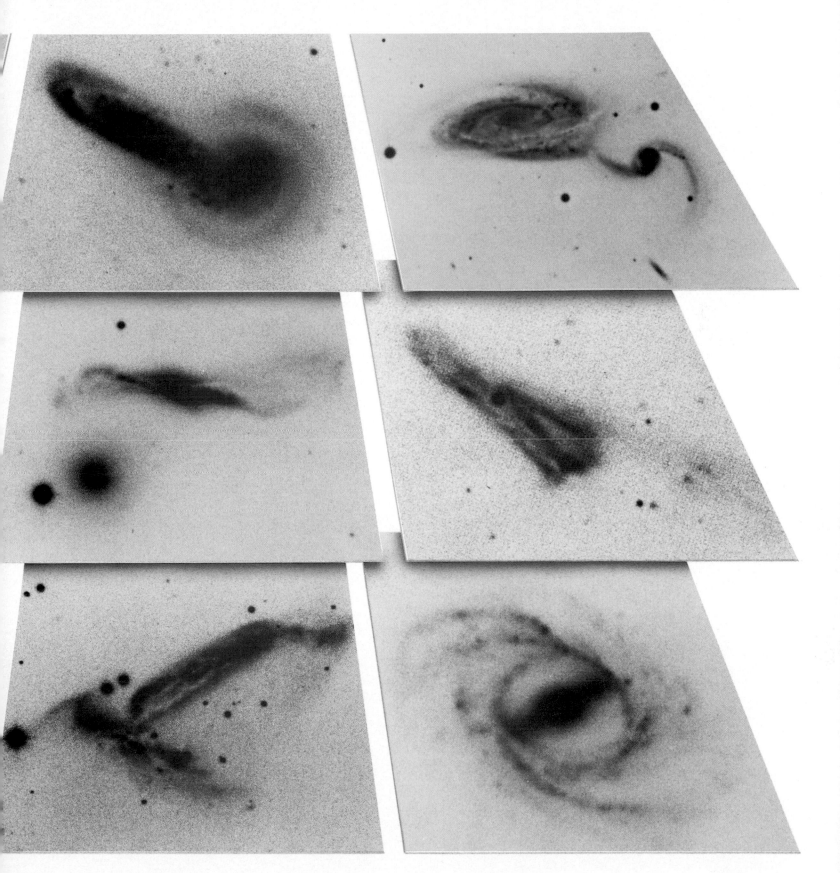

4/ENDURING MYSTERIES

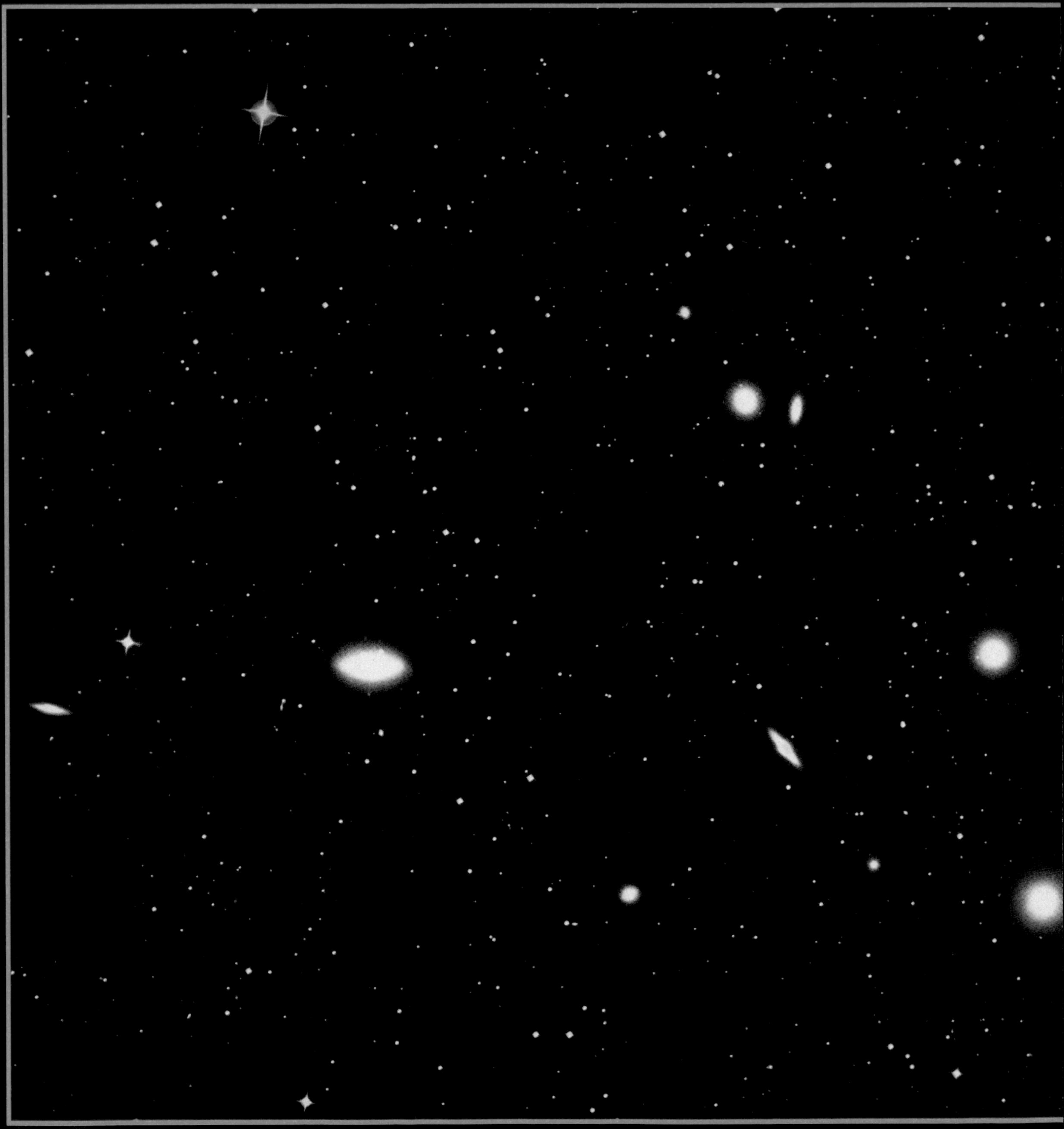

A big barred spiral galaxy *(top right)* dominates several elliptical systems *(center)* in a galactic swarm known as the Fornax cluster. Stars in the foreground reside in the Milky Way.

utting the University of Arizona's brand-new infrared detector through a trial run in the spring of 1987, graduate student Richard Elston pointed it at nothing in particular and began to record the faint traces of heat from deep space. The instrument, developed for military use by the aerospace firm Rockwell International and only recently declassified, crammed 4,096 individual heat sensors onto a microchip smaller than the eraser on the end of a pencil. It was thousands of times more sensitive than previous infrared devices. Elston had little idea what it might turn up.

A few hours later, having rechecked his data three times to be sure of the results, the twenty-seven-year-old astronomer raced down the hall with astonishing news: He had stumbled upon two fuzzy objects that seemed to be more remote than anything else known. Flaring with huge amounts of infrared radiation, they matched the theoretical description of protogalaxies—star systems in the making.

Elston's objects were not the first candidates for protogalactic status, but they were evidently by far the youngest. Their radiation had begun its journey toward Earth 17 billion years ago.

Theorists depict a protogalaxy as a giant cloud of hydrogen, most of it ionized, or electrically charged, by the fierce light of a smattering of newborn stars. The hot young stars should be very luminous at infrared wavelengths; the vast billows of gas, in turn, should radiate brightly in the ultraviolet part of the spectrum, emitting far more total energy than the small stellar population. In 1986 the characteristic ultraviolet radiation had led astronomers to two protogalaxies, at an apparent distance of 12 billion light-years. But those discoveries had been the result of a decade-long search. Elston's chance find at infrared wavelengths suggested that, as he put it, the primeval galaxies "must be everywhere," a thick spawn of star stuff glimpsed at the edges of time.

Whether that is in fact the case has not yet been confirmed, but the intellectual stakes are considerable. None of the accepted models for the evolution of the universe show galaxies forming so early. If the age of Elston's objects is verified and nascent galaxies are determined to be commonplace at that date, adjustments will have to be made in the thinking of cosmologists—astronomers who study the nature of the universe as a whole.

Cosmological convictions have often been reshaped by new observations.

In the 1920s they underwent a revolution. Before Edwin Hubble showed that the universe is expanding, astronomers had largely been content to think of the cosmos as an unchanging collection of gas and stars, without beginning or end in time or space. Once astronomers accepted the Hubble expansion, they began to look for explanations. A logical place to start was with the evolution of galaxies, a subject Hubble himself had skirted in devising his galactic classification scheme *(pages 32-33)*.

EVOLVING IDEAS

Hubble's primary intention had been simply to sort regular galaxies by their recognizable characteristics. But some astronomers suggested that his diagram of the range of galactic shapes actually traced a progression from infant galaxies to old. According to this view, stellar systems might begin their lives as ellipticals, mature to become spirals, and then finally fall apart in senescence as irregulars. From the outset, however, the idea rested on shaky theoretical ground. Spiral galaxies have distinctive features—a flattened disk and spiral arms, and in some instances a central bar. In order for such structures to have evolved from an essentially featureless parent elliptical, marked changes would have had to occur in the orbits of the stars making up the elliptical. But calculations showed that the gravitational interactions needed to achieve the stellar reshuffling would take too long: The universe was not old enough for the present multiplicity of galactic forms to have evolved in this fashion.

By the early 1950s, astronomers concerned with the origin of galaxies often found themselves drafted to one side or the other of an ongoing controversy over the origin of the universe. The debate, between cosmological concepts known as the Big Bang and Steady State theories, reached far beyond the insular astronomical community to catch the interest of a fascinated public.

The Big Bang posited a time billions of years in the past when an incredibly dense concentration of energy and mass, its origin unknown, exploded to form all the matter that now makes up the universe. First put forward in the 1920s by Belgian cosmologist and priest Georges Lemaître, the theory found a vigorous champion in George Gamow, a Russian-born American astrophysicist. Gamow worked out details of nuclear fusion in the Big Bang during the 1940s while he was a professor at George Washington University in Washington, D.C. Capitalizing on his flair for explaining complex concepts to lay audiences, Gamow popularized the theory in his 1952 book, *The Creation of the Universe.*

Until 250 million years after the Big Bang, Gamow maintained, matter took the form of a thin gas, spread evenly throughout space. According to a generally accepted principle of astrophysics, any such gas would, under the influence of its own gravitation, break up into giant clouds whose size would be determined by the density and temperature of the gas. When Gamow plugged his theoretical numbers for these factors into the applicable formula, he found that a typical post-Big Bang gas cloud would have the same mass as an average-size galaxy of today.

These protogalaxies, as Gamow called the gas clouds, would have had an average diameter of about 40,000 light-years and a mass equal to about 200 million Suns. Still moving apart under the impetus of the Big Bang, they would have been, in Gamow's words, "cold, dark and chaotic." But plenty of heat and light would eventually appear as each cloud began to condense and break up into myriad stars through the same process that had formed the clouds themselves. The shape of the resulting galaxy would be determined by the size and rotational speed of the initial gas cloud (*pages 130-131*).

The Big Bang concept, and its attendant explanation of the early history of galaxies, gained widespread but not universal acceptance. One flaw was its failure to explain the origin of the primeval kernel of energy and matter. In the late

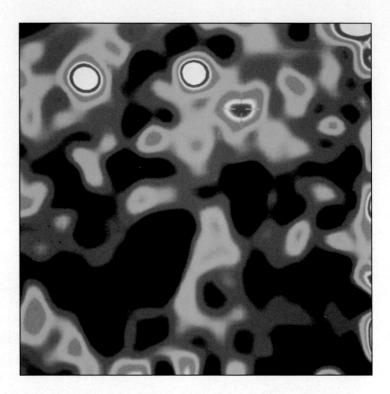

Cross hairs mark a possible protogalaxy, detected by a powerful infrared sensor in 1987. A burst of star formation in the young system generates infrared radiation so intense that it can be measured from a distance of 17 billion light-years. In the false colors of this computer-generated image, the red protogalaxy is only slightly dimmer than two white stars, despite its being about 100,000 times as far away.

1940s, a number of British astronomers proposed the Steady State theory as an alternative. Superficially, Steady State resembled the old, pre-Hubble notion of an unchanging cosmos, but the new theory made adjustments to accommodate the Hubble expansion.

At the forefront of the Steady State theorists was a brash young Cambridge University astronomer and mathematician named Fred Hoyle. Hoyle, barely into his thirties at the time, did not as a rule shy away from absolute statements. The general properties of the universe remain constant, he maintained; it has always been, and will always be, very much as it is now. To account for the expansion discerned by Hubble, the Steady State theory offered a rather magical solution. As the galaxies recede from each other, new matter, in the form of hydrogen atoms, is created in the space between them; ultimately, this matter coalesces into new galaxies. Thus, the average density of any region of space remains the same, although its actual contents are always changing. The theory left unexplained the mechanism by which the new matter would be created, and Hoyle acknowledged there was no practical way to prove the theory. "Obviously we cannot test it in the laboratory," he wrote, "unless we were to find some way to speed up the creation of matter artificially—for the rate of creation, according to theory, is negligible in terrestrial terms." In a volume of space the size of the average physics laboratory, Hoyle figured, hydrogen atoms would materialize at the rate of one every 1,000 years.

The presumed transformation of this tenuous gas into galaxies followed a somewhat different course than the one envisioned by the Big Bang theory. In the Steady State universe, existing galaxies move through space, propagating new stellar systems as they pass. The thin haze of intergalactic hydrogen concentrates in the gravitational wake of a passing galaxy, producing a cloud with enough self-gravitation to condense into a protogalaxy. The protogalaxy, in turn, contracts further to spawn stars and smaller gas clouds.

Hoyle demonstrated a real genius for publicity. He explained his ideas on a widely heard radio series and then published an enormously popular book, *The Nature of the Universe,* in 1950. His efforts won wide acceptance for the theory among the general public, if not among astronomers, who largely held to the Big Bang. Over the course of more than a decade, proponents of the two schools published books and articles, marshaling their evidence and refining the theories.

For neither side was the subject of galaxy formation more than a supporting player in the drama of the creation of the universe itself. Theorists in both camps seemed to agree that, for one reason or another, galaxies had condensed from clouds of gas. Therefore it was something of a shock in 1958 when a prestigious Soviet astronomer proposed an altogether different model—one where galaxies are formed not by the contraction of tenuous gases but from the explosion of dense objects.

Victor Ambartsumian was a veteran theoretician. By the time he graduated from Leningrad University in 1928 at the age of twenty, he had published ten papers on astrophysics and mathematics. It was his belief that observation took precedence over speculation in the development of theory, and he often castigated cosmologists for proposing schemes that could not be verified by available data. In 1946, at the peak of his productive career, Ambartsumian founded the Byurakan Astrophysical Observatory in his native Armenia. The following year he reported a discovery that would form the basis of his own theory of galactic origins.

STELLAR ANOMALIES

Within the Milky Way, Ambartsumian had found what he called stellar associations, groups of ten to a thousand young stars apparently born together in the same region of the galaxy. Although the stars were close enough to each other to be clearly related, he determined that, unlike the stars in open and globular clusters, they were moving too fast for their gravity to hold them together permanently. To Ambartsumian, this trend toward the eventual disintegration of an association indicated that the stars in such a group were fragmentary remnants of the explosion of a small, dense protostar of undetermined origin.

Ambartsumian saw similarities between stellar associations and other cosmic phenomena, such as the ejection of shells of gas and dust from dying suns, and the gradual breakup of the orbits of some widely separated binary stars (stars that circle around one another). In fact, he proposed, similar dispersive processes might be at work everywhere as the fundamental dynamic of the universe.

The discovery of radio galaxies in the 1950s provided Ambartsumian with more ammunition. Contrary to the opinions of such luminaries as Walter

Baade and Rudolph Minkowski, who hypothesized that powerful radio beacons like Cygnus A were the result of a collision between two galaxies, Ambartsumian proposed a single source for the outpouring of such huge amounts of energy. Radio galaxies, he said, "are systems in which ejections of tremendous scale from the nucleus have taken place. As a consequence of such ejections, clouds of high-energy particles are being formed." Noting that there was no known process that could cause such enormous outbursts, Ambartsumian hypothesized the existence of a superdense object erupting at the core of a radio galaxy. He offered the same explanation for quasars; in the tortured, laser-bright cores of these cosmic powerhouses, he saw the explosive galaxy-creation process at work.

Ambartsumian extended his ideas about radio galaxies to more ordinary stellar systems. There too, he said, material is hurled from the nuclei, to form such features as the spiral arms. In sum, he traced the birth of galaxies, and perhaps even that of entire clusters of galaxies, to the violent eruption of compact, supermassive forebears.

A LUKEWARM RECEPTION

Although Ambartsumian's concept of a universe dominated by the evolution of objects from dense matter to more diffuse states generally agreed with the precepts of the Big Bang, its model for the origin of individual galaxies was radically different. When he presented his ideas to a conference of astronomers that was conducted in Brussels in 1958, the reaction was thoughtfully skeptical. Dutch radio astronomer Jan Oort was generally not persuaded by Ambartsumian's galactic theories. "I feel some doubt," he said, "whether there are sufficiently compelling phenomena in the world of galaxies to justify the adoption of such a revolutionary idea as the fission of galaxies."

As it was, Ambartsumian took no part in the larger cosmological controversy. That debate effectively came to an end during the 1960s with the discovery of some apparently conclusive evidence in support of the Big Bang theory. One of the hypothetical by-products of the original cosmic explosion is a weak but pervasive electromagnetic radiation at very short wavelengths— a sort of radio echo of the Big Bang. In 1964 radio astronomers Arno Penzias and Robert Wilson of Bell Telephone Laboratories in New Jersey discovered a faint signal that seemed to come with equal strength from every direction. Its characteristics matched the predicted emanation from

George Abell puts the finishing touches on the Palomar Sky Survey in May 1956. The lower chart records the dates when each of 879 regions was first photographed; the upper chart indicates the delivery of prime-quality plates. At the bottom of each are unfilled spaces, representing the extreme southern skies, not visible from Palomar Mountain.

the Big Bang, and there simply was no other satisfactory explanation for it.

The so-called background radiation rapidly converted most doubters to belief in the Big Bang itself, but it did not square with at least one of the Big Bang's presumed consequences. Events in the earliest epoch after the momentous instant are assumed to have determined the distribution of matter throughout the expanding cosmos. To explain the large-scale structures seen in the universe today—most obviously, galaxies—Big Bang proponents suggest that within the first few minutes there were variations in the concentration of mass from region to region. Called density fluctuations, these variations would cause matter to clump together into ever-greater concentrations. The problem was to reconcile the apparent evenness of the early expansion, as indicated by the steady background radiation, with the observed large-scale structures: A perfectly smooth cosmic explosion would have produced only an increasingly rarefied gas cloud.

The theory George Gamow had worked out in the 1940s did not present a great discrepancy. It postulated only slight density fluctuations, just enough to lead to galaxies. Most astronomers believed that, except for a few anomalous groups, galaxies were evenly spread throughout space.

But galactic distribution was something of a mystery. Paradoxically, the giant telescopes that peer deep into space also act as blinders, preventing a broad picture of the heavens. A telescope like Mount Wilson's 100-inch reflector has such a small field of view that it would take thousands of years to photograph the entire sky. Consequently, astronomers' visions of the universe were based on narrow samples of the sky. By extrapolating from these observations, they arrived at a model of the cosmos as a more or less random scattering of galaxies, interspersed with a few areas where clusters appeared. The group of which the Milky Way was a part—including also the Andromeda galaxy, the Magellanic Clouds, and about thirty other star systems—was considered to be an exception, not the rule.

This notion had begun to come under fire as early as the 1930s, when Harlow Shapley, then at Harvard, conducted a detailed survey of several areas of the sky and showed that galaxies tend to form groups of various sizes. Still, by the end of the 1940s, scarcely three dozen such clusters had been discovered, not enough to force a revision of the commonly held view of large-scale structure in the universe. It remained for a photographic survey of the sky, organized by Rudolph Minkowski and completed in 1956, to confirm beyond any doubt the existence of widespread clustering.

TURNING A NEW EYE ON THE SKY

The survey used a brand-new forty-eight-inch Schmidt telescope at Palomar Observatory. This wide-angle device could cover a sector of the sky hundreds of times broader than the field of a 100-inch telescope. But even with the Big Schmidt, as the survey astronomers dubbed it, the task was an arduous one. Over a period of nearly seven years, the researchers exposed some 3,000 plates to create a high-quality photographic record of the entire visible sky. With

Earth

In four cones of space whose contents have been observed and analyzed from Earth *(center)*, galaxies form clusters, which in turn form larger structures called superclusters. Each cone contains parts of superclusters that appear to be linked with each other in a weblike pattern. The richest clusters—those that are the brightest and most populous—are portrayed in white; the other clusters are shown in descending order of richness as yellow and orange.

A WEBWORK OF SUPERCLUSTERS

A combination of painstaking observation and high-speed computing produces a three-dimensional map of the universe *(left)* in which galaxies—the largest single objects in the cosmos—are found clumped together in clusters and superclusters. The study of these large-scale groupings helps astronomers probe events in the early history of the universe.

The first step in charting the cosmic structure is to determine the two-dimensional coordinates of galaxies and galaxy clusters as they appear in a small sector of the sky. To add the third dimension, the map makers take red-shift readings that indicate how far the bodies are from Earth. The slowness of this process (typically, a good reading for just one galaxy requires about an hour on a large telescope) limits the scope of such surveys to small volumes of the universe.

The positions of the clusters are then analyzed by a computer running a special program. The computer randomly selects one cluster from all those being considered and makes it the center of a spherical volume of space with a radius of about 100 million light-years. Any other cluster that appears within the sphere is considered to be linked to the first cluster as part of a single formation. The computer then puts the second cluster at the center of a sphere and repeats the process. This continues until all the space containing the clusters under study has been examined. Regions where many clusters are found to have near neighbors are considered to be possible superclusters.

To ensure that the patterns detected by the computer are not just coincidental groupings in an otherwise disorderly cosmos, researchers use the same program to analyze a simulated sector of sky with randomly scattered clusters. It appears that there is less than one chance in a million that the structures are the result of chance. Vast superclusters almost certainly reflect an underlying structure in the universe.

an eye sharp enough to photograph a candle flame at a distance of 10,000 miles, the Big Schmidt captured the images of millions of galaxies. On some plates, 50,000 galaxies appeared in an area of the sky no bigger than the bowl of the Big Dipper.

The principal observer for the Palomar survey was a twenty-nine-year-old graduate student named George Abell. Between nights working on the survey to support himself and his growing family, Abell also recorded and analyzed groups of galaxies for his Ph.D. dissertation—a huge undertaking that his professors at the California Institute of Technology had initially discouraged as too difficult for a lone, relatively inexperienced astronomer. Abell amazed them by finishing his degree soon after the completion of the survey, and in 1958 he published a list of 2,712 "rich clusters," concentrations of hundreds or thousands of bright galaxies. The Abell Catalog, as it came to be known, was quickly recognized by astronomers as indispensable: The reference work provided a complete count of rich clusters visible from Palomar up to a distance of three billion light-years.

AN ALLY WITH EVIDENCE

As he studied the distribution of these aggregations, Abell found that they tended to clump together in still larger formations, a phenomenon he called second-order clustering. Many astronomers, reluctant to admit that matter could be so unevenly distributed in the cosmos, were slow to acknowledge Abell's findings. But critical support came from the work of French astronomer Gérard de Vaucouleurs, who pored over data from many sources in studying large-scale structure in the vicinity of the Milky Way. Several years earlier, in 1953, de Vaucouleurs had presented evidence for a "local supergalaxy," a flattened cluster of perhaps tens of thousands of galaxies that spanned about 40 million light-years and was a few million light-years thick. De Vaucouleurs placed the center of this supergalaxy near a cluster of galaxies in the direction of the constellation Virgo. With further study, he concluded by the end of the 1970s that the aggregation was even larger than he had first thought. The Local Supercluster, as he then named it, had a diameter of 160 to 240 million light-years and a mass equal to a million billion Suns.

As additional evidence accumulated, the existence of clusters and superclusters became indisputable, and cosmologists were faced with the need to stretch the Big Bang theory to accommodate the new findings. In the 1970s, they drew up two different lines of attack on the problem. Spearheading one approach was James Peebles of Princeton University, who argued that the large-scale structure of the universe had developed, in effect, from the bottom up. In Peebles's opinion, the galaxies had formed first and congregated to make the clusters later.

Opposing this was the view espoused by Yakov Zel'dovich of the Soviet Academy of Sciences: Any density fluctuations with masses as small as the average galaxy's would have been wiped out by the extreme heat of the infant cosmos. Only very large-scale irregularities, he maintained—irregularities as massive as today's biggest clusters—could have survived to seed the universe as it cooled. Thus, in Zel'dovich's top-down scenario, the clusters condensed first, with galaxies forming afterward.

These two theories produced quite different predictions of how galaxies and clusters would be distributed throughout space. Peebles's model called for the clusters and their member galaxies to be scattered quite randomly. Zel'dovich's calculations, in contrast, implied that over large distances the universe should resemble a piece of very bubbly Swiss cheese, with galaxies confined to thin sheets and filaments. Between these sheets and filaments would be huge dark voids.

By the 1980s a raft of observations seemed to indicate that Zel'dovich's model came closer than Peebles's to describing the distribution of galaxies. Using a variety of techniques, including X-ray detectors as well as optical and radio telescopes, astronomers plotted enormous structures in the universe. In 1987 forty-four-year-old astrophysicist Brent Tully of the University of Hawaii used a supercomputer to plot the relative positions of previously recorded clusters and superclusters. Studying the maps produced by the computer, Tully concluded that de Vaucouleurs's Local Supercluster was in fact part of a vast complex of superclusters filling 10 percent of the observed

Vera Rubin, here surrounded by her collection of antique globes, established in the early 1980s that even the most luminous stellar systems contain great amounts of dark matter, evidenced by the rotation rates of a disk galaxy's stars and gas.

universe. Composed of millions of galaxies, the complex appeared to be one billion light-years long and 150 million light-years wide—more than 100 times larger than any previously known structure. Furthermore, Tully suggested, there were signs of at least four other systems of a similar scale.

Not even Zel'dovich had predicted a universe as lumpy as that described by Tully. A cosmological model that could produce such vast structures would have to include large density fluctuations in the moments after the Big Bang. The catch, of course, is that the resulting uneven expansion should also be reflected in irregularities in the background radiation—which is in fact extremely smooth.

Zel'dovich's theory also stumbles over the great age of galaxies. As the last elements of the system to take shape—after the condensation of superclusters, then clusters—galaxies should not have begun to form until about 10 billion years ago. But judging by the age of globular clusters, the Milky Way itself must be at least 14 billion years old. The discovery of protogalaxies in the late 1980s widened the gap between observation and hypothesis. No theory could encompass galaxy formation 17 billion years ago as well as huge concentrations of galaxies in the present. The enigma of large-scale structure continues to defy solution.

A RIDDLE IN THE DARK

The cosmological picture was further muddled in the 1970s and 1980s by the discovery of previously undetected matter in galaxies. To understand the gravitational interaction of galaxies and clusters, theorists must make assumptions about the objects' masses. Not surprisingly, they base their estimates on what they can see: the luminous stars and gas clouds that show up on photographic plates or on infrared and ultraviolet detectors. But the new findings indicate that visible matter is only a small fraction of the total in the universe. Like the legendary blind men, theorists had been describing an elephant while clinging to the tip of its tail.

The research team that first detected the unseen material was led by Vera Rubin, an astronomer at the Carnegie Institution in Washington, D.C. She had already triggered one controversy in the 1960s when she reported observations that showed the Milky Way moving at a high velocity with respect to distant galaxies. No one could explain the origin of such motion, and many astronomers simply refused to believe Rubin's results. The heat of the debate—which has never reached a definitive resolution—impelled Rubin to look for, as she put it, "a safe problem, where no one would bother me." Her choice seemed innocuous: She would take spectrographic readings of spiral galaxies to determine their rotation patterns, then study the relationship between those interior dynamics and the shapes of the galaxies.

The spectrograph she employed was a remarkable device designed by Kenneth Ford, whose reputation as an instrument maker had originally lured Rubin to the Carnegie Institution. With it, she could get precise measurements of the spectra of gas clouds in the faint outer reaches of distant galaxies. Over a six-year span at observatories in Arizona and Chile, Rubin and her colleagues observed sixty carefully selected galaxies at a rate of two or three per night.

INTIMATIONS OF THE INVISIBLE

The results astonished Rubin and created a new stir among astronomers. She had expected the spectrogram of a galaxy to show rotation velocities that rose rapidly from the center and then fell off toward the outer part of the disk *(pages 42-43)*. This would indicate that the mass was distributed in proportion to the luminous matter, with the heaviest concentration near the center. Instead, a pattern very different from the one anticipated emerged on the first night of observation. Near the center of each galaxy, velocities rose as expected with increased distance, but instead of falling, they either leveled off or kept rising all the way to the galaxy's visible edge. "It was phenomenal," Rubin later recalled. "After one night, with four spectrograms, we knew something strange was going on."

Subsequent observations confirmed the first night's results: Regardless of their size or type, all of the selected galaxies showed similar rotation patterns. The only plausible explanation was that their mass, far from crowding toward the center as the luminous matter seemed to indicate, was more

A tapestry of light and dark reveals the large-scale texture of the universe on a map of galaxies within a billion light-years of Earth. Divided into a million squares, each shaded according to the number of galaxies it contains (from black for none to white for ten or more), the map shows galaxies in clusters, filaments, and broad clouds across the sky.

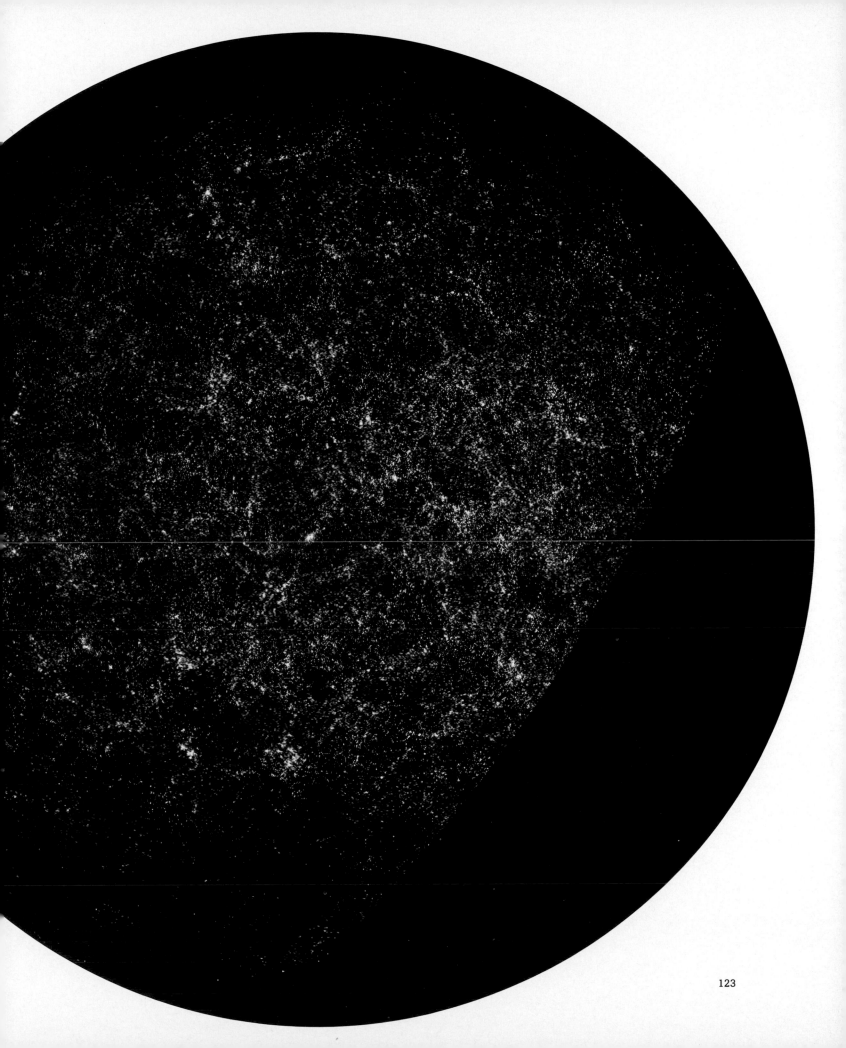

evenly spread throughout—and well beyond—the visible circumference of the galaxy. As much as 90 percent of the matter in each galaxy was invisible, detectable only by its powerful gravitational influence.

The spectrographic evidence of this dark matter, as it came to be called, was irrefutable, and astronomers quickly accepted Rubin's conclusions. No one could explain just what the invisible mass was—and Rubin did not try, remarking that it could be anything from subatomic particles to black holes. The mystery remains to this day, exciting considerable interest among astronomers. The interest is understandable, for dark matter may hold the key to one of the biggest of cosmological questions: the ultimate fate of the universe.

There seem to be two choices. Either the Hubble expansion could continue forever, or gravitational forces could slow the fleeing galaxies and finally stop them, thus starting a process of contraction that would persist until the universe reached something like the original state before the Big Bang. The issue hinges on the average density of matter throughout space; if it exceeds a certain value—the equivalent of about three hydrogen atoms per cubic meter—the cosmos is doomed to eventual contraction. And dark matter in galaxies increases the estimated density to nearly the level required to halt expansion. If more such matter exists in the dark voids between superclusters, for example, the universe may be headed irrevocably for a Big Crunch.

Like so many recent discoveries, dark matter reminds astronomers that galaxies are rich in complexity and surprise. Still, as Vera Rubin put it, "Science progresses best when observations force us to alter our preconceptions." The story of galaxies has been a story of tumbling preconceptions, and recent work in the field suggests that astronomy can expect more of the same.

THE GENESIS OF A STAR SYSTEM

The seeds of modern galaxies were sown when time, space, energy, and matter exploded out of the Big Bang some 12 to 20 billion years ago. Physicists can only speculate about the dynamics and distribution of primordial matter, but one thing is virtually certain: After millions of years, the universe was suffused with large, irregular aggregations of gas or stars held loosely together by the force of gravity. It was from these cosmic chrysalises—called protogalaxies—that modern galaxies emerged. Just how protogalaxies themselves were formed is still under debate; two of the most popular models for protogalactic birth are shown in the following pages.

Scientists understand better the transformation from protogalaxy to final form. In a grand metamorphosis, the protogalaxy contracts and gives birth to wave after wave of new stars, culminating in the ellipticals and disk-shaped galaxies that float through the cosmos.

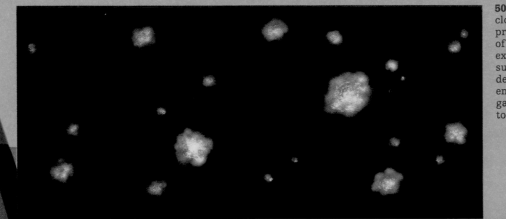

500,000 years after the Big Bang. Ho[t] clouds of hydrogen and helium—the primordial matter—dot this early view of the universe. No other elements yet exist in significant quantities. At a sunlike temperature of a few thousan[d] degrees Kelvin, the universe has coole[d] enough for these denser pockets of gas—some the size of dwarf galaxies— to clump together.

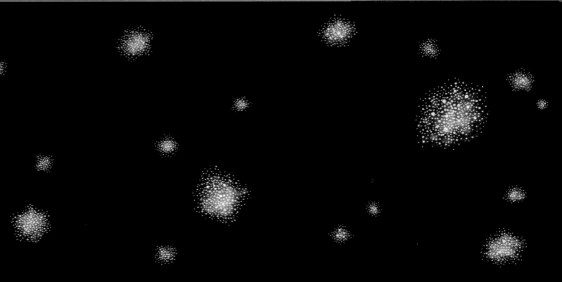

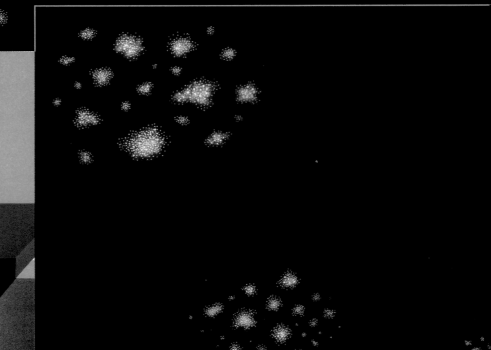

SPACE EXPANDS

126

A PROCESS OF CONGREGATION

Most theorists assume that the Big Bang did not scatter the makings of the cosmos evenly through space but, rather, that the explosion had tiny flaws in its symmetry. As the universe expanded, these small disparities were magnified, resulting in an unequal distribution of matter. Thus, at a very early point in cosmic evolution, some regions of space contained slightly denser concentrations of primordial hydrogen and helium gas, the universe's first elements. According to the most common view of galaxy formation, clusters of stars developed from these pockets of denser matter.

Once the original star groups were born, the theory continues, they floated at random through the early universe, meeting other clusters along the way. Bigger clusters brought smaller ones under their gravitational sway, creating still larger structures that swept up even more clusters as they moved through space. This process culminated in massive stellar assemblies held together in a loose-knit gravitational web: protogalaxies.

Because this theory proposes that large-scale structures such as galaxies and galaxy clusters arose from smaller groups of stars, it is often described as the "bottom-up" picture. Though still highly regarded, the theory has at least one shortcoming: It does not explain the recent discovery of superclusters—immense groupings of galaxy clusters spanning hundreds of millions of light-years. The gradual clumping of matter into ever larger aggregations proceeds too slowly to allow for the formation of superclusters in the time since the Big Bang.

One billion years after the Big Bang. Further condensation has caused bursts of star formation within each cloud, using up essentially all of the cloud's original store of hydrogen and helium. The space between the new star clusters is virtually empty.

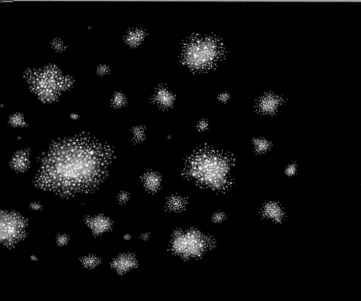

Two to three billion years after the Big Bang. Star clusters have merged to form distinct, yet loosely bound, protogalaxies, separated by the otherwise unpopulated void. Gravity holds these irregular bodies intact; within them, the stars age and the most massive begin to explode as supernovae. The explosions enrich the interstellar medium with gas that includes elements such as carbon and oxygen, created in the nuclear furnaces within stars.

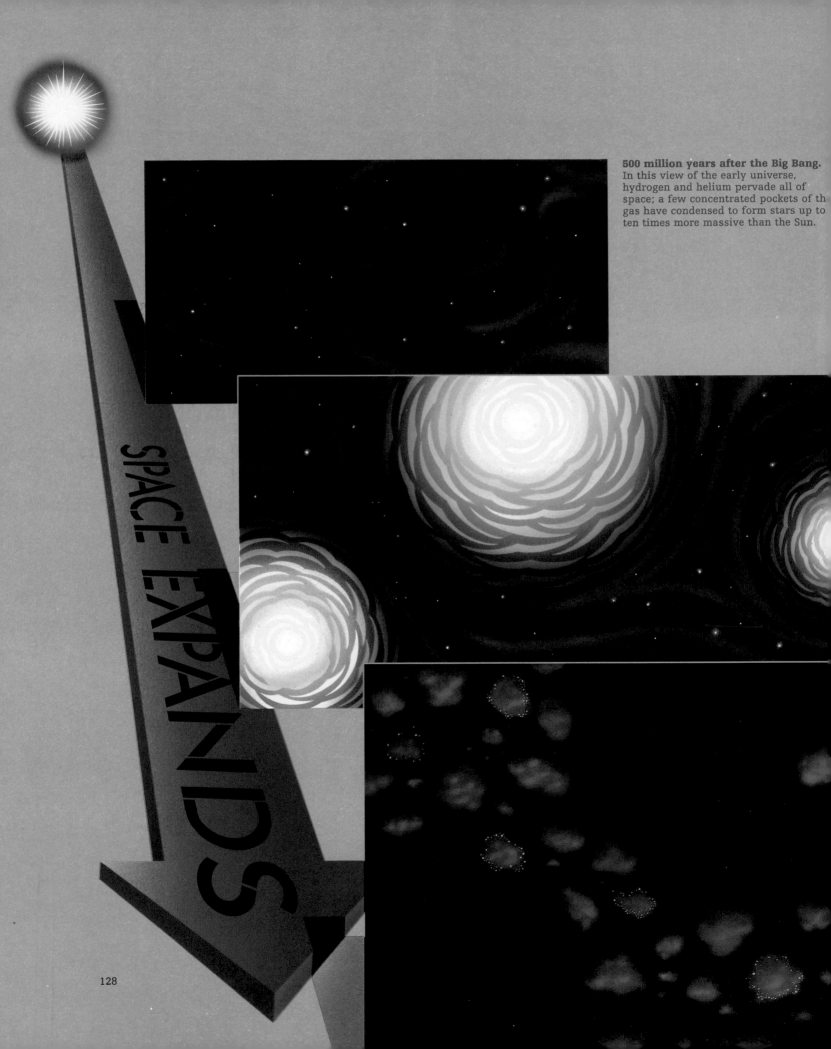

500 million years after the Big Bang. In this view of the early universe, hydrogen and helium pervade all of space; a few concentrated pockets of th gas have condensed to form stars up to ten times more massive than the Sun.

SPACE EXPANDS

Birth by Cosmic Blasts

Another explanation for the birth of protogalaxies invokes sweeping shock waves in what could be called the cosmic explosions theory. Unlike the bottom-up model, which begins with the relatively rapid formation of star clusters from dense clumps of gas, this theory assumes that the primordial gas was more evenly distributed and required some external force to start the evolution from gas to protogalaxy.

In one scenario, a small number of stars formed from the densest regions of gas in the early cosmos. The most massive of them blasted outward as supernovae, creating cataclysmic shock waves that pushed the surrounding gas into thick clouds. The clouds then condensed into stars.

A more novel approach places so-called cosmic strings at the center of the turbulence that spawns protogalaxies. Predicted by the exotic physics of Big Bang studies, strings have never been observed. In theory, they are threads left over from the first moments of the universe, invisibly thin, vibrating tubes containing vast amounts of energy. Since strings would not have expanded with the rest of the universe, they would be unbelievably dense and massive, weighing 10 million billion tons per inch, though only 10^{-30} centimeters thick. More important, they would oscillate at speeds near that of light, losing so much energy that eventually they would dissolve. This energy emission could create shock waves that would then compress surrounding gas, causing star clusters and protogalaxies to form.

One billion years after the Big Bang. Reaching the end of their comparatively short lives, some of the first massive stars blast outward in supernova explosions. These cataclysms create a host of new elements and give rise to shock waves that compress the intervening gas. Successive wave fronts create ropy, high-density regions where gas clouds are pushed together and low-density regions where the gas is swept away. The compressed gas begins to collapse into stars.

Five billion years after the Big Bang. The dense, elongated clouds have collapsed into collections of stars and gas. Larger assemblies have become protogalaxies; smaller ones are now star clusters. Together they form a vast, curving chain around the emptier spaces that have been sculpted by earlier supernovae. Only a few star clusters remain in isolation.

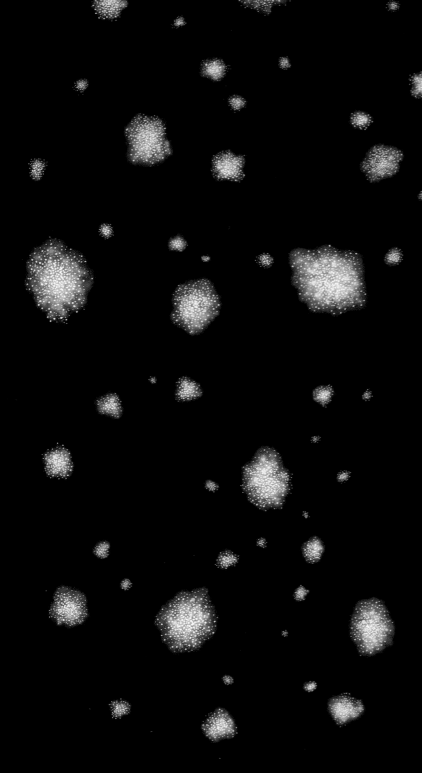

A protogalaxy begins to collapse into its final form. Irregular in shape and clumpy in structure, it consists of a loose network of gaseous star clusters in random motion.

Contracting further, the galaxy becomes more symmetrical. Some stars die as supernovae, ejecting gas enriched with heavy elements into the space between adjacent star clusters and creating new stars as shock waves condense local gases. The shrinking galaxy's final form is still undetermined.

THE MAKING OF AN ELLIPTICAL

Once a protogalaxy forms and begins to contract under the relentless force of gravity, its evolutionary pace accelerates. As the gas and stars of the protogalaxy's clusters fall toward each other, the largest gather around a gravitational center to form a galactic nucleus. Gravitational interaction with other protogalaxies causes the protogalaxy to spin slowly, but as matter falls inward, the rate of rotation picks up.

The continuing collapse of the stars toward the nucleus spurs several bursts of rapid star formation in the clusters. When some of the ensuing stars die as supernovae, the gas produced in the explosions is enriched with elements heavier than hydrogen and helium. As this gas falls toward the nucleus, it forms new stars, which evolve and eventually eject still more gas laden with heavy elements—elements crucial to the development of planets and the subsequent evolution of life.

In the final phases of galaxy formation, the galaxy's shape—elliptical or spiral—will be determined primarily by the rate at which it makes stars. A lumpy distribution of gas may give rise to more rapid star formation, using up the protogalaxy's available gas and producing an elliptical galaxy whose stars are smoothly distributed from a dense core outward. Spin velocity also affects overall shape: The faster a galaxy rotates, the flatter its final form will be.

The galaxy stops collapsing inward when the available gas is used up and stars reach a stable orbit around the galactic nucleus. Its evolution complete, the newborn elliptical is a fraction of its original size, condensed to a shining orb a tenth to a thirtieth the size of the protogalaxy from which it sprang.

The larger star clusters in the protogalaxy begin to gravitate toward its center, forming a bright nucleus. A dimmer halo of diffuse gas and globular star clusters surrounds the center. As the random motion of its gas and stars yields to a more organized orbital motion, the protogalaxy's spin becomes more obvious.

An elliptical galaxy is born as the final collapse of halo gases and globular clusters toward the nucleus coincides with very rapid star formation, effectively consuming all of the elliptical's free gas. The resulting galaxy is brightest at its core, dimming gradually toward an almost imperceptible edge.

THE SLOW GROWTH OF A DISK

Spiral galaxies begin to differ from ellipticals only when they enter their adolescence, after their nuclei begin to form. Once the more massive stars have gathered in the galaxy's central bulge, star production in would-be disk galaxies slows greatly, leaving quantities of unused hydrogen or helium gas. Astronomers believe that gas in the nucleus and outer halo of a young spiral may exert enough pressure on the thin intervening gas to suppress star generation there.

The final evolution of a spiral galaxy begins when the protogalaxy collapses in the same way shown on pages 130-131, its stars circling around a nucleus. Unlike an elliptical galaxy, the early spiral contains more diffuse gas in the halo around the nucleus, and thus it forms stars more slowly.

Diffuse gas clouds in random motion about the nucleus collide and merge in a process that will ultimately create the central disk of the spiral galaxy. The upward and downward movements of the whirling patches of gas cancel out, and the clouds retain only their orbital motion around the nucleus, as represented by the merging arrows.

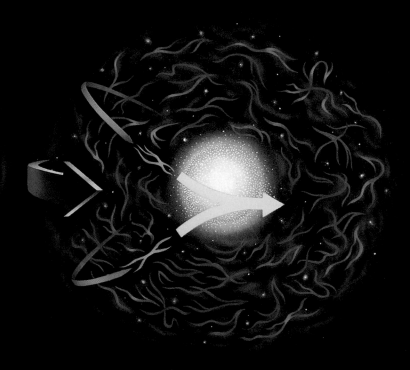

Because this intervening gas is unable to form stars, it continues to swirl in various directions around the nucleus. Patches collide and merge in their travels, and their random paths tend to converge in the familiar disk shape, a shape that represents the most stable compromise between the pull of gravity and the energy of the gas's original motion about the nucleus. (The origin of the disk's spiral pattern is still unknown, however.)

Elliptical galaxies may be considered complete when their free gas has condensed into stars and the stellar birthrate has plummeted. Spiral galaxies, on the other hand, mature more gradually. Gas in the disk condenses slowly into stars, and the last remnants of free gas continue to fall into the galactic plane even after more than 10 billion years. Indeed, many astronomers believe that the disk of one typical spiral—the Milky Way—is still growing.

Star formation proceeds slowly, and gas clouds gradually precipitate out of the halo into an orbit along the equatorial plane of the galaxy. A partial disk develops, forming from the nucleus outward; a faint spiral structure begins to appear. As the gas clears out of the halo, globular star clusters become visible.

A fully formed spiral galaxy takes shape as stars form in the disk and diffuse gas finishes its precipitation. Globular clusters remaining from the protogalaxy hover around the disk and the central bulge in a vast sphere.

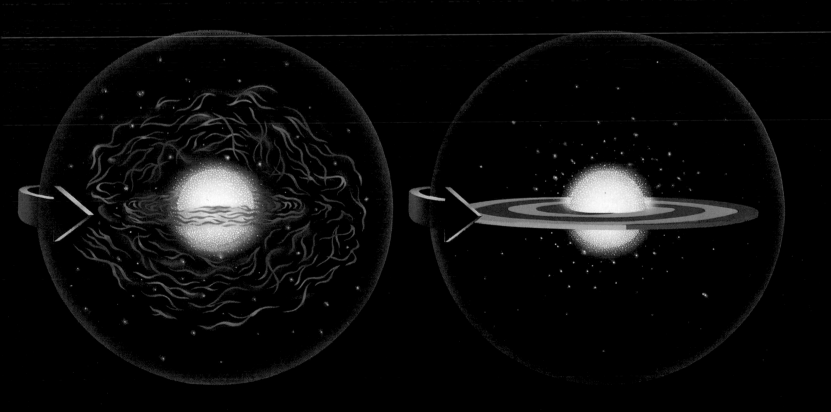

GLOSSARY

Absorption line: a dark band at a particular frequency on a spectrum that forms when substances between a light source and an observer absorb light of that frequency. Different substances produce characteristic patterns of absorption lines.

Accretion disk: a disk formed from gases and other materials that are drawn in by a compact body, such as a black hole, at the disk's center.

Active galaxy: a galaxy with a highly energetic nucleus.

Angular diameter: an object's width on the celestial sphere, measured in degrees of arc. The Moon's angular diameter is just over half a degree.

Angular momentum: a measure of an object's inertia, or state of motion, about an axis of rotation.

Antenna: in radio astronomy, the component of a radio telescope that converts received radio waves into electrical signals.

Astronomical Unit (AU): a distance unit, often used within the Solar System, equal to just over 93 million miles, the approximate distance between the Earth and the Sun.

Astrophysics: the study of the physics of astronomical objects and processes, made possible by the development of spectroscopy and photography in the nineteenth century.

Atom: the smallest component of a chemical element that retains the properties associated with that element. Atoms are composed of protons, neutrons, and electrons.

Background radiation: a steady emission of electromagnetic radiation detectable from all directions.

Barred spiral galaxy: a disk galaxy in which the spiral arms begin at the ends of a central bar rather than at the nucleus.

Binary stars: a pair of stars formed at the same time and orbiting around their mutual center of gravity. Between one-half and two-thirds of the stars in the Sun's neighborhood are members of binary systems.

Big Bang: according to a widely accepted theory, the primeval moment, 15 to 20 billion years ago, when the universe began expanding from a single point.

BL Lacertae object: an N-galaxy with a particularly bright nucleus that changes rapidly in luminosity; named for BL Lacertae, the first such galaxy to be identified.

Black hole: in theory, an extremely compact body with such great gravitational force that no radiation can escape from it.

Blue shift: a Doppler effect seen when a radiating source approaches the observer; the received wavelengths are compressed so that any absorption and emission lines move from their expected frequencies toward the shorter-wavelength blue end of the spectrum. *See* red shift.

Bridge: a filament of stars stretching between two galaxies that may result from the galaxies' interaction.

Brightness: the amount of light received from an object; a combined result of its actual luminosity, its distance, and any light absorption by interstellar dust or gas.

Bulge: a large sphere of stars at the center of a spiral galaxy.

Catalog: a published list of astronomical objects and their precise positions.

Celestial coordinates: a pair of numbers designating an object's location on the celestial sphere. One coordinate, declination, is a north-south value similar to latitude; the other, right ascension, is similar to longitude.

Celestial sphere: the apparent sphere of sky that surrounds the Earth; used by astronomers as a convention for specifying the location of a celestial object.

Celsius: a scientific temperature scale in which 0° is the freezing point and 100° the boiling point of water.

Cepheid variable: a star that changes regularly in luminosity over a set period of days or weeks.

Charge-coupled device (CCD): an electronic array of electromagnetic radiation detectors, usually positioned at a telescope's focus.

Cluster (of galaxies): a gravitationally bound system of galaxies, ranging in number from a few dozen to several thousand. The Milky Way belongs to the Local Group cluster.

Collision: the gravitational interaction of galaxies at very close range, sometimes resulting in a merger.

Constellation: originally a pattern of stars named for an object, animal, or person but now more commonly the area of sky assigned to that pattern. Every astronomical object is in a specific constellation.

Continuous spectrum: a spectrum consisting of all wavelengths in a given range, without absorption or emission lines.

Cosmic ray: an atomic nucleus or other charged particle moving at close to the speed of light.

Cosmic string: according to theory, a flaw developed during the early expansion of the universe after the Big Bang that retains many of the properties of that period.

Cosmology: the study of the universe as a whole, including its large-scale structure and movements, origin, and ultimate fate.

Dark matter: a form of matter whose existence is deduced from its observed gravitational effects but that is not directly observable at any wavelength.

Density wave: a moving pattern of compression and rarefaction. Density waves are thought to maintain spiral structure in galaxies.

Differential rotation: rotation in which components at different distances from the center orbit at different rates, as opposed to rigid, solid-body rotation.

Dish: a colloquial term for a radio telescope design in which a large bowl-shaped structure focuses received radio waves on a central point; may also refer to a large optical mirror.

Disk: the pancake-shaped component of a spiral galaxy, made up of stars, gas, and dust; often incorporates bright spiral arms.

Disk galaxy: any galaxy with a disk, including spiral and barred spiral galaxies. Astronomers using computer simulations to study galactic structure and interactions often employ disk galaxies with no spiral features.

Doppler effect: a wave phenomenon in which waves appear to compress as their source approaches the observer or stretch out as the source recedes from the observer. *See* red shift, blue shift.

Double radio source: an active galaxy consisting of two radio lobes flanking an elliptical galaxy.

Dust: interstellar material consisting of widely separated microscopic particles of ice-coated rock.

Dynamical friction: a dragging force created by the combined gravitational effects of several bodies.

Electromagnetic radiation: radiation consisting of periodically varying electric and magnetic fields that vibrate perpendicularly to each other and travel through space at the speed of light.

Electromagnetic spectrum: the array, in order of frequency or wavelength, of electromagnetic radiation, from low-frequency, long-wavelength radio waves to high-frequency, short-wavelength gamma rays.

Electron: a negatively charged particle that normally orbits an atom's nucleus but may exist in isolation.

Elliptical galaxy: a galaxy shaped like a flattened sphere, with no discernible features and no disk.

Emission line: a bright band at a particular frequency on a spectrum, emitted directly by the source and indicating by its frequency a chemical constituent of that source.

Expanding universe: the expansion of space itself as a result of the Big Bang.

Extragalactic: located outside the Milky Way galaxy.

Fahrenheit: a nonastronomical temperature scale in which 32° Fahrenheit is equivalent to 0° Celsius and 212° Fahrenheit is equivalent to 100° Celsius.

Filament: a threadlike structure of gases, stars, or galaxies.

Frequency: the number of oscillations per second of an electromagnetic wave. *See* wavelength.

Galaxy: a system of stars, gas, and dust that contains from millions to hundreds of billions of stars.

Gamma rays: the most energetic form of electromagnetic radiation, with the highest frequency and shortest wavelength. Because the Earth's atmosphere absorbs most radiation at this end of the spectrum, gamma ray astronomy must be performed from space.

Gas: a component of most galaxies, accounting for a significant fraction of a galaxy's mass. Most gas in the universe is a form of hydrogen.

Globular cluster: a spherical system of up to a few million stars that normally orbits the center of a galaxy.

Gravity: the mutual attraction of separate masses; a fundamental force of nature.

Guide star: a star tracked by a guide telescope during a long photographic exposure to ensure that the main telescope is accurately aligned.

HI, HII: *see* hydrogen.

Halo: in a spiral galaxy, a spherical volume hundreds of thousands of light-years across, centered on the bulge, and defined by the orbits of extremely old stars and globular clusters.

Hubble constant: an estimate of the rate at which the universe is expanding. Because values for the Hubble constant vary from 50 to 100 kilometers (30 to 60 miles) per second per megaparsec, astronomers usually specify which value they have used in converting a galaxy's red shift to its distance from Earth.

Hydrogen: the most common detectable element in the universe. It occurs in several forms, including neutral hydrogen (HI), hydrogen atoms with no electrical charge, containing a proton and an electron; ionized hydrogen (HII), positively charged hydrogen atoms from which the electron has been stripped; and molecular hydrogen, two-atom molecules of hydrogen.

Hydrogen line: *see* twenty-one-centimeter line.

Infrared: a band of electromagnetic radiation with a lower frequency and a longer wavelength than visible red light. Most infrared radiation is absorbed by the Earth's atmosphere, but certain wavelengths can be detected from Earth.

Intensity: the amount of radiation received from an object. Optical astronomers prefer the term *brightness*.

Interacting galaxies: galaxies that are or have been at close enough range to affect each other gravitationally.

Interferometer: in radio astronomy, an arrangement of two or a few radio telescopes used as one to locate sources of radio emission. With components separated by a distance several times the wavelength of radio waves, an interferometer takes advantage of the natural interference of radio waves as they arrive at the system from different directions.

Interstellar absorption: the conversion of starlight into heat by interstellar dust, thus reducing the light received on Earth.

Interstellar absorption line: an absorption line resulting from light's passage through interstellar dust and gas rather than from the outer layers of the radiating object.

Inverse-square law: the fact that, under perfectly transparent conditions, an object's brightness is inversely proportional to the square of its distance. A star at two light-years is four times dimmer than it would be at one.

Ion: an atom that has lost or gained one or more electrons. In comparison, a neutral atom has an equal number of electrons and protons, giving the atom a zero net electrical charge. A positive ion, such as HII, has fewer electrons than the neutral atom; a negative ion has more.

Irregular galaxy: a galaxy with an amorphous shape, neither spiral nor elliptical. *See* peculiar galaxy.

Jet: a thin, high-speed plasma stream ejected from the center of a radio source.

Kapteyn Universe: an early-twentieth-century model of the galaxy that placed the Sun near the center; named after its originator, Dutch astronomer Jacobus Kapteyn.

Kelvin: an absolute temperature scale that uses Celsius degrees but sets 0 at absolute zero, or about -273° Celsius.

Light: the part of the electromagnetic spectrum that is visible.

Light-year: an astronomical distance unit equal to the distance light travels in a vacuum in one year, almost six trillion miles.

Local Group: the cluster of thirty or more galaxies to which the Milky Way belongs.

Luminosity: an object's total energy output, usually measured in ergs per second.

Lunar occultation: the Moon's passage in front of an object, which allows precise measures of the object's size, location, and features.

Magellanic Clouds: the Large and Small Magellanic Clouds, which are the two galaxies closest to Earth, visible from the Southern Hemisphere.

Magellanic Stream: a trail of hydrogen gas extending from the Magellanic Clouds toward the Milky Way's south pole.

Magnetic field line: an indication of the direction and strength of a magnetic field at a particular location.

Magnitude: a designation of an object's brightness or luminosity relative to that of other objects; apparent magnitude refers to observed brightness, absolute magnitude to an object's hypothetical brightness at a standard distance of about 32.6 light-years. By convention, absolute magnitude is used in astronomy as an indication of an object's actual luminosity.

Mass: a measure of the total amount of material in an object, determined either by its gravity or by its tendency to stay in motion, if in motion, or at rest, if at rest.

Merger: a collision between two or more galaxies that results in their stars combining to form a single galaxy.

Messier catalog: an eighteenth-century catalog compiled

by French astronomer Charles Messier to keep track of nebulae; includes several galaxies, designated by their position in the catalog. For example, M31, the Andromeda galaxy, is the thirty-first entry.

Milky Way: the Earth's galaxy, a giant spiral of at least one hundred billion stars. The Sun is two-thirds of the way out from the Milky Way's center.

Molecule: the smallest component of a chemical that retains the chemical's properties. A molecule consists of one or more atoms bonded together.

N-galaxy: an active elliptical galaxy with a small, very bright nucleus that varies in luminosity.

Nebula: a cloud of interstellar dust or gas, in some cases a supernova remnant or a shell ejected by a star. Nebulae once included all soft-edged objects, such as galaxies and globular clusters; this usage is no longer correct.

Neutrino: a chargeless particle with little or no mass.

New General Catalogue **(NGC):** a catalog of star clusters, galaxies, and nebulae published in 1888. Many galaxies are named for their position in the catalog; for example, NGC 5178 is the 5,178th entry.

Neutron: an uncharged particle with a mass similar to a proton's; normally found in an atom's nucleus.

Nonthermal emission: electromagnetic radiation produced by any of several processes, including synchrotron radiation. The characteristic pattern of a nonthermal emission is that it increases in intensity as frequency increases.

Noise: meaningless random changes in radiation that tend to obscure a specific signal.

Nova: a star that exhibits a sudden, temporary increase in brightness thousands of times its normal appearance.

Nucleus: the core of a galaxy; also, the center of an atom, composed of protons and neutrons and orbited by electrons.

Open cluster: a loose grouping of stars, ranging in number from a few dozen to a few thousand, found in the plane of a galaxy; contrasted with globular cluster.

Orbit: the path of an object revolving around another object or point.

Orbital decay: the outcome when a body moves too slowly to sustain its orbit; the orbit becomes smaller and smaller until the orbiting and orbited bodies collide.

Parallax (stellar): a star's apparent motion on the celestial sphere over a six-month period. Measured in seconds of arc, it is used to determine a star's distance; the larger the parallax, the nearer the star.

Parsec: an astronomical distance unit equal to approximately 3.26 light-years. A star would be one parsec from Earth if it exhibited a parallax of one second of arc. The dimensions of galaxies are often given in kiloparsecs, or thousands of parsecs; larger distances are given in megaparsecs, or millions of parsecs.

Peculiar galaxy: a galaxy that has a more defined shape than a typically amorphous irregular but is not clearly a spiral or an elliptical. Many peculiar galaxies are thought to result from a galactic interaction.

Period-luminosity relation: a correlation pertaining to Cepheid variable stars that links a Cepheid's absolute magnitude to the length of its period, or cycle of brightening and dimming (the longer the period, the brighter the star); a critical tool for estimating extragalactic distances. By comparing a Cepheid's absolute magnitude with its apparent magnitude, astronomers can calculate the star's distance.

Photometry: the measurement of an object's brightness, or apparent magnitude, by means of an instrument that counts the number of photons detected in a time period.

Photon: a unit of electromagnetic energy associated with a specific wavelength.

Planetary nebula: a shell of gas, ejected from a red giant star; appears planetlike in a low-powered telescope.

Plasma: a gaslike association of ionized particles that responds collectively to electric and magnetic fields. Because plasma particles do not interact the way particles of ordinary gas do, plasma is considered a fourth state of matter, along with solid, liquid, and gas.

Polar-ring galaxy: a peculiar galaxy with an elongated central body encircled by a pole-to-pole ring of stars at right angles to the plane of the central galaxy.

Positron: a particle similar to an electron but positively charged.

Proper motion: the movement of an object across the celestial sphere, reflecting its actual motion relative to the Sun. In contrast, stellar parallax is a yearly cycle relative to the Earth that does not produce a net change of position.

Protogalaxy: a roughly spherical hydrogen cloud from which a galaxy forms; about thirty times the size of a mature galaxy.

Proton: a positively charged particle with about 2,000 times the mass of an electron; normally found in an atom's nucleus.

Quasar: an object with a quasi-stellar appearance, now believed to be a tiny, extremely bright, and extremely distant active galaxy that emits intense radiation at most wavelengths.

Radio: the least energetic form of electromagnetic radiation, with the lowest frequency and longest wavelength.

Radio astronomy: the observation and study of radio waves produced by astronomical objects.

Radio galaxy: a source of radio emissions associated with an optically identified galaxy; normally an active galaxy.

Radio lobe: a vast expanse of plasma highly luminous at radio or other wavelengths and emitted by the core of an extragalactic radio source.

Radio telescope: an instrument for studying astronomical objects at radio wavelengths.

Red shift: a Doppler effect seen when a radiating source recedes from the observer. The received wavelengths lengthen so that any absorption and emission lines move from their expected frequencies toward the long-wavelength red end of the spectrum. *See* blue shift.

Resolution: the degree to which details in an image can be separated, or resolved. The resolving power of a telescope is usually proportional to the diameter of its mirror.

Ring galaxy: a peculiar galaxy with a central core surrounded by a ring in the same plane; thought to result from the collision of a disk galaxy and another, smaller galaxy.

Seyfert galaxy: an active disk galaxy with a very bright, starlike nucleus; first identified by American astronomer Carl Seyfert in 1943.

Shock wave: in astronomy, a sudden discontinuity in the flow of a gas, liquid, or plasma characterized by abrupt increases in temperature, pressure, and velocity.

Solar mass: a stellar mass unit equal to the Sun's mass, about two billion trillion trillion grams.

Solar system: the Sun and the planets, asteroids, and other bodies that orbit the Sun; more generally, any star together with the objects that orbit it.

Spectral type: a star's classification, based on its spectrum, according to an established system organized by surface temperature and mass.

Spectrogram: a photograph of an astronomical spectrum.

Spectroscopy: the study of spectra.

Spectrum: the array of colors or frequencies obtained by dispersing light, as through a prism; often banded with absorption or emission lines.

Spiral arm: a dense region in the disk of a spiral galaxy, containing very young stars and ionized hydrogen clouds.

Spiral galaxy: a disk galaxy with bright stars and ionized gas clouds that form a pattern of two or more spiral arms curving out from the galaxy's center.

Steady State: an alternate theory to the Big Bang, not widely accepted, stipulating that the universe has always existed, and will always exist, in a state similar to that of the present.

Stellar association: a loose cluster of stars, believed to have formed together.

Stellar population: a broad classification of stars based on their age, galactic location, chemical composition, spectra, and motions.

Sunspot: a temporary, cool feature on the Sun's surface that appears dark by contrast with its surroundings.

Supercluster: a large association of clusters of galaxies.

Supermassive black hole: a hypothetical object at the center of some galaxies, with a mass equivalent to several hundred million stars.

Supernova: a star explosion that expels all or most of the star's mass and is extremely luminous.

Supernova remnant: an expanding nebula, consisting of the mass ejected by a supernova.

Synchrotron radiation: a type of nonthermal emission generated by electrons and cosmic rays spiraling around magnetic field lines at near light-speed.

Tail: a curved filament of stars extending from a galaxy on the side opposite to its interaction with another galaxy.

Thermal emission: electromagnetic radiation produced by heat-related processes and characterized by an emission pattern that falls in intensity as frequency increases.

***Third Cambridge Catalogue* (3C):** a catalog of radio sources published in 1959.

Tidal force: the force resulting from the varying gravitational attraction exerted by one body on parts of another.

Twenty-one-centimeter line: a spectral line produced by neutral hydrogen at a radio wavelength of just over twenty-one centimeters; the first radio spectral line to be detected.

Ultraviolet: a band of electromagnetic radiation with a higher frequency and shorter wavelength than visible blue light. Most ultraviolet is absorbed by Earth's atmosphere, so ultraviolet astronomy is normally performed in space.

Variable star: a star that changes in luminosity over time. Some variable stars change predictably and repeatedly; others change unpredictably or only once.

Velocity: the speed and direction of motion.

Violent relaxation: a rapid approach to gravitational equilibrium by multiple bodies, such as the stars in interacting galaxies.

Wave: the propagation of a pattern of disturbance.

Wavelength: the distance from crest to crest or trough to trough of an electromagnetic wave; related to frequency (the longer the wavelength, the lower the frequency).

X rays: a band of electromagnetic radiation intermediate in wavelength between ultraviolet radiation and gamma rays. Because X rays are completely absorbed by the atmosphere, X-ray astronomy must be performed in space.

BIBLIOGRAPHY

Books

Abell, George O., David Morrison, and Sidney C. Wolff, *Exploration of the Universe.* Philadelphia: Saunders College Publishing, 1987.

Arp, Halton, *Quasars, Redshifts, and Controversies.* Berkeley, Calif.: Interstellar Media, 1987.

Athanassoula, E., ed., *Internal Kinematics and Dynamics of Galaxies.* Boston: D. Reidel, 1983.

Baker, Robert H., *Astronomy.* New York: D. Van Nostrand, 1964.

Berendzen, Richard, Richard Hart, and Daniel Seeley, *Man Discovers the Galaxies.* New York: Neale Watson Academic Publications, 1976.

Bergamini, David, and the Editors of *Life, The Universe* (Life Nature Library series). New York: Time, 1962.

Berman, Louis, and J. C. Evans, *Exploring the Cosmos.* Boston: Little, Brown, 1986.

Binney, J., J. Kormendy, and S. White, *Morphology and Dynamics of Galaxies.* Ed. by L. Martinet and M. Mayor. Sauverny/Switzerland: Geneva Observatory, 1982.

Biographical Encyclopedia of Scientists. New York: Facts on File, 1981.

Bok, Bart J., and Priscilla F. Bok, *The Milky Way.* Cambridge, Mass.: Harvard University Press, 1981.

Bondi, Hermann, et al., *Rival Theories of Cosmology.* London: Oxford University Press, 1960.

Burnham, Robert, Jr., *Burnham's Celestial Handbook* (3 vols.). New York: Dover, 1978.

Calder, Nigel, *Violent Universe: An Eyewitness Account of the New Astronomy.* New York: Viking Press, 1969.

Campbell-Jones, Simon, ed., *At the Edge of the Universe.* New York: Universe Books, 1983.

Duncan, John Charles, *Astronomy.* New York: Harper & Brothers, 1955.

Edge, David O., and Michael J. Mulkay, *Astronomy Transformed: The Emergence of Radio Astronomy in Britain.* New York: John Wiley & Sons, 1976.

Fall, S. M., and D. Lynden-Bell, eds., *The Structure and Evolution of Normal Galaxies.* New York: Cambridge University Press, 1980.

Ferris, Timothy:
Galaxies. New York: Stewart, Tabori & Chang, 1982.
The Red Limit: The Search for the Edge of the Universe. New York: William Morrow, 1977.

French, Bevan M., and Stephen P. Maran, eds., *A Meeting with the Universe.* Washington, D.C.: National Aeronautics and Space Administration, 1981.

Gillispie, Charles Coulston, ed., *Dictionary of Scientific Biography.* New York: Charles Scribner's Sons, 1981.

Henbest, Nigel:
  The Exploding Universe. New York: Macmillan, 1979.
  Mysteries of the Universe. New York: Van Nostrand
  Reinhold, 1981.
Henbest, Nigel, and Michael Marten, *The New Astronomy.*
  New York: Cambridge University Press, 1983.
Hey, J. S., *The Evolution of Radio Astronomy.* New York:
  Science History Publications, 1973.
Hodge, Paul W., *Galaxies.* Cambridge, Mass.: Harvard
  University Press, 1986.
Hodge, Paul W., comp., *The Universe of Galaxies: Read-
  ings from Scientific American.* New York: W. H.
  Freeman, 1984.
Hubble, Edwin Powell, *The Realm of the Nebulae.* New
  Haven, Conn.: Yale University Press, 1936.
Kaufmann, William J., III:
  Planets and Moons. San Francisco: W. H. Freeman,
  1979.
  Universe. New York: W. H. Freeman, 1985.
Kippenhahn, Rudolf, *100 Billion Suns.* Transl. by Jean
  Steinberg. New York: Basic Books, 1983.
La Cotardière, Philippe de, ed., *Larousse Astronomy.* New
  York: Facts on File, 1987.
Lang, Kenneth R., and Owen Gingerich, eds., *A Source
  Book in Astronomy and Astrophysics, 1900-1975.*
  Cambridge, Mass.: Harvard University Press, 1979.
Laustsen, Svend, Claus Madsen, and Richard M. West,
  Exploring the Southern Sky. New York: Springer-Verlag,
  1987.
Learned, R., *Astronomy through the Telescope.* New York:
  Van Nostrand Reinhold, 1981.
Mallas, John H., and Evered Kreimer, *The Messier Album.*
  Cambridge, Mass.: Sky, 1978.
Mayall, N. U., "Edwin Powell Hubble." In *Biographical
  Memoirs* (Vol. 41). New York: Columbia University
  Press, 1970. (Published for the National Academy of
  Sciences).
Mihalas, Dimitri, and James Binney, *Galactic Astronomy:
  Structure and Kinematics.* San Francisco: W. H.
  Freeman, 1981.
Mitton, Simon, *Exploring the Galaxies.* New York: Charles
  Scribner's Sons, 1976.
Mitton, Simon, ed., *The Cambridge Encyclopedia of
  Astronomy.* London: Jonathan Cape, 1977.
Moore, Patrick, ed., *The International Encyclopedia of
  Astronomy.* New York: Orion Books, 1987.
Morrison, Philip, and Phylis Morrison, *Powers of Ten.*
  New York: Scientific American Books, 1982.
Murdin, Paul, and David Allen, *Catalogue of the Universe.*
  New York: Cambridge University Press, 1979.
Page, Thornton, and Lou Williams Page, eds., *Beyond the
  Milky Way: Galaxies, Quasars, and the New Cosmology.*
  New York: Macmillan, 1969.
Pasachoff, Jay M., and Marc L. Kutner, *University Astron-
  omy.* Philadelphia: W. B. Saunders, 1978.
Preiss, Byron, and Andrew Fraknoi, eds., *The Universe.*
  New York: Bantam Books, 1987.
Ridpath, Ian, ed., *Illustrated Encyclopedia of Astronomy
  and Space.* New York: Thomas Y. Crowell, 1979.
Ronan, Colin A., *Deep Space.* New York: Macmillan, 1982.
Ryle, Martin, N. Kurti, and R. L. F. Boyd, *Search and
  Research.* Ed. by John P. Wilson. London: Mullard,
  1971.

Sandage, Allan, *The Hubble Atlas of Galaxies.* Washing-
  ton, D.C.: Carnegie Institution of Washington, 1961.
Sciama, D. W., *The Unity of the Universe.* Garden City,
  N.Y.: Doubleday, 1959.
Shapley, Harlow, *Through Rugged Ways to the Stars.* New
  York: Charles Scribner's Sons, 1969.
Shipman, Harry L., *Black Holes, Quasars, and the Uni-
  verse.* Boston: Houghton Mifflin, 1980.
Shu, Frank H., *The Physical Universe.* Mill Valley, Calif.:
  University Science Books, 1982.
Smith, Elske v. P., and Kenneth C. Jacobs, *Introductory
  Astronomy and Astrophysics.* Philadelphia: W. B.
  Saunders, 1973.
Smith, Robert, *The Expanding Universe.* New York:
  Cambridge University Press, 1982.
Snow, Theodore P., *Essentials of the Dynamic Universe:
  An Introduction to Astronomy.* St. Paul: West, 1987.
Struve, Otto, Beverly Lynds, and Helen Pillans, *Elementa-
  ry Astronomy.* New York: Oxford University Press,
  1959.
Struve, Otto, and Velta Zebergs, *Astronomy of the Twenti-
  eth Century.* New York: Macmillan, 1962.
Sullivan, W. T., III, ed., *The Early Years of Radio Astrono-
  my: Reflections Fifty Years after Jansky's Discovery.*
  New York: Cambridge University Press, 1984.
Van Woerden, Hugo, Willem N. Brouw, and Henk C. van
  de Hulst, eds., *Oort and the Universe.* Boston:
  D. Reidel, 1980.
Verschuur, Gerrit L., *The Invisible Universe Revealed.*
  New York: Springer-Verlag, 1986.
Waterfield, Reginald L., *A Hundred Years of Astronomy.*
  London: Duckworth, 1938.
Whitney, Charles A., *The Discovery of Our Galaxy.* New
  York: Alfred A. Knopf, 1971.
Zeilik, Michael, and Elske v. P. Smith, *Introductory
  Astronomy and Astrophysics.* New York: Saunders
  College Publishing, 1987.

Periodicals
Abell, George O., "Exploring the Farthest Reaches of
  Space." *National Geographic,* December 1956.
Balick, Bruce, "Quasars with Fuzz." *Mercury,* May/June
  1983.
Bowen, Ira Sprague, "Sky Survey Charts the Universe."
  National Geographic, December 1956.
Boyd, Mary Jo, "When Galaxies Collide." *Science Digest,*
  June 1986.
Burton, W. B., and H. S. Liszt, "The Gas Distribution in
  the Central Region of the Galaxy." *Astrophysical
  Journal,* November 1, 1978.
Carney, Bruce W., "Probing the Outer Galactic Halo."
  Publications of the Astronomical Society of the Pacific,
  November 1984.
Clark, Gail O.:
  "Ancients of the Universe." *Astronomy,* May 1985.
  "Stellar Populations." *Astronomy,* October 1986.
Darling, David, "What Makes a Spiral Galaxy?" *Astrono-
  my,* July 1979.
De Vaucouleurs, Gérard, "The Supergalaxy." *Scientific
  American,* July 1954.
Disney, Michael J., and Philippe Véron, "BL Lacertae
  Objects." *Scientific American,* August 1977.
Eicher, David J., "Observing the Local Group of Galax-

ies." *Astronomy,* November 1984.

Fosbury, R. A. E., and T. G. Hawarden, "A0035, 'the Cartwheel'—a Large Southern Ring Galaxy." *Monthly Notices of the Royal Astronomical Society,* 1977.

Friel, Eileen, "A Symposium on Stellar Populations." *Mercury,* November/December 1984.

Gamow, George:
"The Evolutionary Universe." *Scientific American,* September 1956.
"Modern Cosmology." *Scientific American,* March 1954.

Gingerich, Owen, "Robert Trumpler and the Dustiness of Space." *Sky and Telescope,* September 1985.

Gingerich, Owen, and Barbara Welther, "Harlow Shapley and the Cepheids." *Sky and Telescope,* December 1985.

Goldstein, Alan, "Observing Interacting Galaxies." *Deep Sky,* spring 1987.

Graham Smith, F., and Bernard Lovell, "On the Discovery of Extragalactic Radio Sources." *Journal for the History of Astronomy,* October 1983.

Gray, George W., "A Larger and Older Universe." *Scientific American,* June 1953.

Gribbin, J., "The Structure of the Universe." *New Scientist,* October 29, 1987.

Hartley, Karen, "Ring around the Galaxy." *Science News,* October 31, 1987.

Hirshfeld, Alan, "Inside Dwarf Galaxies." *Sky and Telescope,* April 1980.

Hoyle, Fred, "The Steady-State Universe." *Scientific American,* September 1956.

Hut, Piet, and Gerald Jay Sussman, "Advanced Computing for Science." *Scientific American,* October 1987.

Lemonick, Michael D., "Light at the End of the Cosmos." *Time,* January 25, 1988.

Lesh, Janet Rountree, "Swarms of Stars." *Astronomy,* March 1978.

Liszt, H. S., and W. B. Burton, "The Gas Distribution in the Central Region of the Galaxy." *Astrophysical Journal,* December 15, 1978.

Lo, K. Y., "The Galactic Center: Is It a Massive Black Hole?" *Science,* September 26, 1986.

Lynds, Roger, and Alar Toomre, "On the Interpretation of Ring Galaxies: The Binary Ring System II Hz 4." *Astrophysical Journal,* October 15, 1976.

Maran, Stephen P.:
"Cosmic Collisions." *Natural History,* August 1987.
"Deep in the Heart of the Milky Way." *Natural History,* October 1978.
"The Quasar Controversy Continues." *Natural History,* January 1982.
"Ring Galaxies." *Natural History,* November 1977.
"Where's the Gas?" *Natural History,* August 1985.

Margon, Bruce, "The Origin of the Cosmic X-Ray Background." *Scientific American,* January 1983.

Marschall, Lawrence, "Galactic Coronas." *Astronomy,* November 1982.

Mathewson, D. S., "The Clouds of Magellan." *Scientific American,* April 1985.

Mathewson, D. S., M. N. Cleary, and J. D. Murray, "The Magellanic Stream." *Astrophysical Journal,* June 1, 1974.

Morrison, Nancy D., and Stephen Gregory, "The Nucleus and Nuclear Bulge of Our Galaxy." *Mercury,* July/August 1985.

Murai, Tadayuki, and Mitsuaki Fujimoto, "The Magellanic Stream and the Galaxy with a Massive Halo." *Publications of the Astronomical Society of Japan,* 1980.

Murphy, Jamie, "The Milky Way's Hungry Black Hole." *Time,* June 3, 1985.

"New Theory Proposed on Galactic Evolution." *Astronomy,* November 1981.

Oort, J. H., "The Development of Our Insight into the Structure of the Galaxy between 1920 and 1940." *Education in and History of Modern Astronomy,* August 25, 1972.

Osmer, Patrick S., "Quasars as Probes of the Distant Early Universe." *Scientific American,* February 1982.

Overbye, Dennis:
"Exploring the Edge of the Universe." *Discover,* December 1982.
"The Shadow Universe." *Discover,* May 1985.

Parker, Barry:
"Celestial Pinwheels: The Spiral Galaxies." *Astronomy,* May 1985.
"Discovery of the Expanding Universe." *Sky and Telescope,* September 1986.

Paul, E. Robert:
"The Death of a Research Programme: Kapteyn and the Dutch Astronomical Community." *Journal for the History of Astronomy,* June 1981.
"J. C. Kapteyn and the Early Twentieth-Century Universe." *Journal for the History of Astronomy,* 1986.

Pearson, John F., "The Sizzling Enigmas at the Edge of Space." *Popular Mechanics,* September 1965.

Percy, John R., "Cepheids: Cosmic Yardsticks, Celestial Mysteries?" *Sky and Telescope,* December 1984.

Reddy, Francis, "To Sculpt the Galaxies." *Astronomy,* January 1983.

Rieke, G. H., et al., "Infrared Astronomy after IRAS." *Science,* February 21, 1986.

Rubin, Vera C.:
"Dark Matter in Spiral Galaxies." *Scientific American,* June 1983.
"Women's Work." *Science 86,* July/August 1986.

Schweizer, François, "Colliding and Merging Galaxies." *Science,* January 17, 1986.

Schweizer, François, Bradley C. Whitmore, and Vera C. Rubin, "Colliding and Merging Galaxies. II. S0 Galaxies with Polar Rings." *Astronomical Journal,* July 1983.

Scoville, Nick, and Judith S. Young, "Molecular Clouds, Star Formation and Galactic Structure." *Scientific American,* April 1984.

Seeley, D., and R. Berendzen:
"The Development of Research in Interstellar Absorption, c. 1900-1930" (Part 1). *Journal for the History of Astronomy,* February 1972.
"The Development of Research in Interstellar Absorption, c. 1900-1930: Part 2." *Journal for the History of Astronomy,* June 1972.

Shields, Gregory, "The Chemistry of Galaxies." *Astronomy,* June 1981.

Silk, Joseph, "Formation of the Galaxies." *Sky and Telescope,* December 1986.

Silk, Joseph, et al., "The Large-Scale Structure of the Universe." *Scientific American,* October 1983.

Skinner, G. K., et al., "Hard X-Ray Images of the Galactic Centre." *Nature,* December 10, 1987.

Smith, David H., "Spirals from Order and Chaos." *Sky and Telescope*, August 1987.

Smith, Robert W., "The Great Debate Revisited." *Sky and Telescope*, January 1983.

Strom, Richard G., George K. Miley, and Jan Oort, "Giant Radio Galaxies." *Scientific American*, August 1975.

Thomsen, D. E., "Galaxies in a Primitive State?" *Science News*, January 23, 1988.

Toomre, Alar, and Juri Toomre:
"Galactic Bridges and Tails." *Astrophysical Journal*, December 15, 1972.
"Violent Tides between Galaxies." *Scientific American*, December 1973.

Trefil, James, "From Astronomy to Astrophysics." *Wilson Quarterly*, summer 1987.

Walz-Chojnacki, Greg, "The Great Galaxy in Andromeda." *Odyssey*, November 1987.

Weaver, Harold:
"Steps toward Understanding the Large-Scale Structure of the Milky Way" (Part 1). *Mercury*, September/October 1975.
"Steps toward Understanding the Large-Scale Structure of the Milky Way" (Part 2). *Mercury*, November/

December 1975.
"Steps toward Understanding the Large-Scale Structure of the Milky Way" (Part 3). *Mercury*, January/February 1976.

Weedman, Daniel W., "Seyfert Galaxies, Quasars and Redshifts." *Quarterly Journal of the Royal Astronomical Society*, Vol. 17, 1976.

Weymann, Ray J., "Seyfert Galaxies." *Scientific American*, January 1969.

Whitmore, Bradley C., and Marylin Bell, "IC 4767 (the 'X-Galaxy'): The Missing Link for Understanding Galaxies with Peanut-Shaped Bulges?" *Astrophysical Journal*, January 15, 1988.

Wyckoff, Susan, and Peter A. Wehinger, "Are Quasars Luminous Nuclei of Galaxies?" *Sky and Telescope*, March 1981.

Yusef-Zadeh, Farhad, and Mark Morris:
"Interaction of Thermal and Nonthermal Radio Structures in the Arc Near the Galactic Center." *Astronomical Journal*, November 1987.
"The Linear Filaments of the Radio Arc Near the Galactic Center." *Astrophysical Journal*, November 15, 1987.

INDEX

Numerals in italics indicate an illustration of the subject mentioned.

PICTURE CREDITS

The sources for the illustrations that appear in this book are listed below. Credits from left to right are separated by semicolons, from top to bottom by dashes.
Cover: Art by David Jonason/The Pushpin Group. Front and back endpapers: Art by John Drummond. 1: David Malin, Anglo-Australian Observatory. 2: Anglo-Australian Telescope Board. 3: David Malin, Anglo-Australian Observatory. 4: © California Institute of Technology, 1959. 5: European Southern Observatory, Garching, FDR. 6: © 1987, Dr. James D. Wray, McDonald Observatory. 12-13: European Southern Observatory, Garching, FDR. 14: Computer-generated initial cap by John Drummond, detail from photo appearing on pages 12-13. 16-17: Art by Sam Ward. 20: Art by Damon Hertig. 22: Art by William Hennessy. 26-27: Courtesy Owen Gingerich, Harvard-Smithsonian Center for Astrophysics; Mary Lea Shane Archives, Lick Observatory; AIP Niels Bohr Library; AIP Niels Bohr Library, E. Scott Barr Collection; from *The Astronomical Scrapbook: Skywatchers, Pioneers and Seekers in Astronomy,* by Joseph Ashbrook, published by Sky Publishing Corporation, Cambridge, Mass.; AIP Niels Bohr Library, W. F. Meggars Collection; Harvard College Observatory; Royal Danish Academy of Science and Letters. Art by Stephen Wagner. 28-29: Lowell Observatory photograph; Mary Lea Shane Archives, Lick Observatory; from the Collection of Owen Gingerich, Harvard-Smithsonian Center for Astrophysics; courtesy the Archives, California Institute of Technology. Art by Stephen Wagner. 30-31: Werts Studio, Inc., courtesy Mount Wilson and Las Campanas Observatories, Carnegie Institution of Washington. 32-33: Official Naval Observatory photograph; National Optical Astronomy Observatories; © California Institute of Technology—National Optical Astronomy Observatories—Yerkes Observatory

(2); Palomar Observatory photograph; California Institute of Technology (2). Background diagram from *The Realm of the Nebulae,* by Edwin Hubble, published by Yale University Press, © 1936, 1982. 35-43: Art by Alfred Kamajian. 44-45: European Southern Observatory, Garching, FDR. 46: Computer-generated initial cap by John Drummond, detail from photo appearing on pages 44-45. 50-51: Inset art by William Hennessy; National Optical Astronomy Observatories—Art by William Hennessy (2). 52-53: Mount Wilson and Las Campanas Observatories, Carnegie Institution of Washington; Yerkes Observatory/University of Chicago; photo by Dorothy Davis Locanthi, courtesy AIP Niels Bohr Library, Dorothy Davis Locanthi Collection; courtesy Professor Doctor Jan Hendrik Oort; Mary Lea Shane Archives, Lick Observatory. Art by Stephen Wagner. 54-55: Courtesy David Jansky; courtesy National Radio Astronomy Observatory and Grote Reber; James Stokley; courtesy Professor Hendrik Cornelis van de Hulst; Yerkes Observatory; CSIRO/Division of Radio Physics; Beverly Lynds, courtesy Sky Publishing Corporation; courtesy Frank Shu. Art by Stephen Wagner. 57: Courtesy AT&T. 60-61: Art by Sam Ward. 62-63: National Optical Astronomy Observatories—art by Damon Hertig. 64-67: Art by Damon Hertig. 69-71: Art by David Jonason/The Pushpin Group. 72-73: Art by David Jonason/The Pushpin Group; H. S. Liszt, National Radio Astronomy Observatory. 74-77: Art by David Jonason/The Pushpin Group. 78-79: Anthony Readhead-Caltech, courtesy *Discover* Magazine. 80: Computer-generated initial cap by John Drummond, detail from photo appearing on pages 78-79. 81: California Institute of Technology Archives. 82: Nuffield Radio Astronomy Laboratories, Jodrell Bank, Macclesfield, Cheshire. 84-85: © Board of Regents, University of Hawaii; courtesy NRAO/AUI; Gaylin Laughlin/IPAC, California Institute of Technology; Ralph Bolin and Ted Stecher/Science Photo Library; C. Jones and C. Stern, Center for Astrophysics. 86-87: R. A. M. Walterbos and R. C. Kennicutt, courtesy National Optical Astronomy Observatories and Leiden Observatory; Nuffield Radio Astronomy Laboratories, Jodrell Bank, Cheshire; R. A. M. Walterbos, E. Brinks, and W. W. Shane, from *Astronomy and Astrophysics,* Supplement Series, Vol. 61, 1985. 88: Geoff Chester. 90-91: Cyril Hazard; Allan Sandage; Maarten Schmidt, California Institute of Technology. 92-93: J. R. Eyerman/*Time* magazine (2). 94-95: Art by Peter Sawyer. 96: Halton Arp, Max Planck Institut für Astrophysik, Garching, FDR. 100-101: Art by Matt McMullen. 102-103: Art by Matt McMullen; photograph by François Schweizer. 104-105: Photograph by François Schweizer—art by Matt McMullen. 106-107: Art by Matt McMullen; photograph by Halton Arp. 108-109: Art by Matt McMullen; photograph by François Schweizer; Larry Sherer, copied from *Atlas of Peculiar Galaxies,* by Halton Arp, California Institute of Technology, Pasadena, 1966 (9). 110-111: Royal Observatory, Edinburgh, © 1984. 112: Computer-generated initial cap by John Drummond, detail from photo appearing on pages 110-111. 114: Richard Elston, George Riehe, Marcia Rieke. 116: J. Baylor Roberts, © 1956 National Geographic Society. 118-119: Art by Alfred Kamajian. 121: Mark Godfrey. 122-123: M. Seldner, B. Siebers, E. J. Groth, and P. J. E. Peebles, 1977, *Astronomical Journal,* Vol. 82, p. 249. 125-133: Art by Stephen Wagner.

ACKNOWLEDGMENTS

The index for this book was prepared by Barbara L. Klein. The editors also wish to thank Joan Adams, Librarian, Jodrell Bank, London; Roy Allen, Sydney; Halton Arp, Richard Black, Mount Wilson and Las Campanas Observatories, Pasadena, Calif.; Ralph Bohlin, Space Telescope Science Institute, Baltimore; Ronald Bracewell, Stanford, Calif.; Robert A. Brown, Space Telescope Science Institute, Baltimore; Paul Ceruzzi, National Air and Space Museum, Washington, D.C.; Steve Charlton, National Optical Astronomy Observatories, Tucson, Ariz.; Brenda Corbin, United States Naval Observatory, Washington, D.C.; Gérard de Vaucouleurs, University of Texas, Austin; Richard Dreiser, Yerkes Observatory, Williams Bay, Wis.; Michael Fitchett, Space Telescope Science Institute, Baltimore; Richard Fleming, National Radio Astronomy Observatory, Green Bank, W.Va.; John Gallagher, Lowell Observatory, Flagstaff, Ariz.; H. G. Gorwin, University of Texas, Austin; Jesse Greenstein, California Institute of Technology, Pasadena; Sherwood Harrington, Astronomical Society of the Pacific, San Francisco; Michael Hauser, NASA Goddard Space Flight Center, Greenbelt, Md.; Robert Havlen, National Radio Astronomy Observatory, Charlottesville, Va.; Cyril Hazard, Institute of Astronomy, Cambridge, England; Peter D. Hingley, Librarian, Royal Astronomical Society, London; John Huchra, Harvard-Smithsonian Center for Astrophysics, Cambridge, Mass.; David Jansky, Little Silver, N.J.; Christine Jones, Harvard-Smithsonian Center for Astrophysics, Cambridge, Mass.; William Kaufmann, University of California, San Diego; William Keel, University of Alabama, Tuscaloosa; Gaylin Laughlin, California Institute of Technology, Pasadena; Ruth Lombardi, Bell Telephone Laboratories, Short Hills, N.J.; Richard Lovelace, Cornell University, Ithaca, N.Y.; Sir Bernard Lovell, London; David Malin, Anglo-Australian Observatory, Sydney; Stephen Maran, James Marsh, NASA Goddard Space Flight Center, Greenbelt, Md.; Thomas Matthews, University of Maryland, College Park; George Miley, Space Telescope Science Institute, Baltimore; Tony Moller, United States Naval Observatory, Washington, D.C.; Anne-Marie de Narbonne, Librarian, Observatoire de Paris, Paris; Patrick S. Osmer, California Institute of Technology, Pasadena; Henry Palmer, Hampshire, England; Anthony Readhead, Roger Romani, California Institute of Technology, Pasadena; Allan Sandage, Mount Wilson Observatory, Carnegie Institution of Washington, Pasadena, Calif.; Dorothy Schaumberg, University of California, Santa Cruz; Maarten Schmidt, California Institute of Technology, Pasadena; Richard Schmidt, United States Naval Observatory, Washington, D.C.; Frank Shu, University of California, Berkeley; A. Richard Thompson, National Radio Astronomy Observatory, Charlottesville, Va.; Alar Toomre, Massachusetts Institute of Technology, Cambridge; Rene Walterbos, University of California, Berkeley; Althea Washington, NASA, Audio-Visual Materials, Washington, D.C.; Margaret Weems, Charlottesville, Va.; Richard M. West, European Southern Observatory, Munich; Farhad Yusef-Zadeh, NASA Goddard Space Flight Center, Greenbelt, Md.

Time-Life Books Inc.
is a wholly owned subsidiary of
TIME INCORPORATED

FOUNDER: Henry R. Luce 1898-1967

Editor-in-Chief: Jason McManus
Chairman and Chief Executive Officer:
J. Richard Munro
President and Chief Operating Officer:
N. J. Nicholas Jr.
Editorial Director: Ray Cave
Executive Vice President, Books: Kelso F. Sutton
Vice President, Books: George Artandi

TIME-LIFE BOOKS INC.
EDITOR: George Constable
Executive Editor: Ellen Phillips
Director of Design: Louis Klein
Director of Editorial Resources: Phyllis K. Wise
Editorial Board: Russell B. Adams, Jr., Dale M.
Brown, Roberta Conlan, Thomas H. Flaherty, Lee
Hassig, Donia Ann Steele, Rosalind Stubenberg,
Henry Woodhead
Director of Photography and Research:
John Conrad Weiser
Assistant Director of Editorial Resources:
Elise Ritter Gibson

PRESIDENT: Christopher T. Linen
Chief Operating Officer: John M. Fahey, Jr.
Senior Vice Presidents: Robert M. DeSena, James
L. Mercer, Paul R. Stewart
Vice Presidents: Stephen L. Bair, Ralph J. Cuomo,
Neal Goff, Stephen L. Goldstein, Juanita T.
James, Hallett Johnson III, Carol Kaplan, Susan
J. Maruyama, Robert H. Smith, Joseph J. Ward
Director of Production Services: Robert J.
Passantino

Editorial Operations
Copy Chief: Diane Ullius
Production: Celia Beattie
Library: Louise D. Forstall

Correspondents: Elisabeth Kraemer-Singh (Bonn);
Maria Vincenza Aloisi (Paris); Ann Natanson
(Rome). Valuable assistance was also provided by
Judy Aspinall, Christine Hinze (London); Dick
Berry (Tokyo); Wibo van de Linde (Amsterdam);
John Dunn (Melbourne); Barbara Gevene Hertz
(Copenhagen); Mary Johnson (Stockholm); Angie
Lemmer (Bonn); Christina Lieberman (New York).

VOYAGE THROUGH THE UNIVERSE

SERIES DIRECTOR: Roberta Conlan
Series Administrator: Judith W. Shanks

Editorial Staff for *Galaxies*
Designer: Ellen Robling
Associate Editor: Blaine Marshall (pictures)
Text Editors: Peter Pocock (principal), Pat Daniels
Researchers: Esther Ferington (principal), Patti H.
Cass, Tina S. McDowell
Assistant Designer: Barbara McClenahan
Editorial Assistant: Alice T. Marcellus
Copy Coordinator: Darcie Conner Johnston
Picture Coordinator: Sharon D. Doggett

Special Contributors: Joseph Alper, Ken Croswell,
David Darling, Bonnie Gordon, Joanne Heckman,
John I. Merritt, Greg Mock, Diana Morgan, Steve
Olson, Carl Posey, Mark Washburn (text); Vilasini
Balakrishnan, Vasilis Basios, Sanjoy Ghosh, John
Harney, Janet Heller, Eugenia Scharf, Tom
Sodroski, Wayne Stewart, Julie Trudeau
(research).

CONSULTANTS

KIRK D. BORNE works at the Space Telescope Science Institute in Baltimore, where he specializes in research on interacting and colliding galaxies.

GEOFFREY BURBIDGE is a professor of physics at the University of California, San Diego. He has published several studies in astrophysics and cosmology, including a seminal paper in 1957 on the synthesis of elements in stars and, in 1967, one of the first comprehensive works on quasars.

JACK O. BURNS is director of the Institute for Astrophysics at the University of New Mexico. He is also a consultant in space plasma physics at the Los Alamos National Laboratory. His research interests include observational cosmology and radio observations of active galaxies and quasars.

GEOFFREY R. CHESTER has been a member of the staff of the Smithsonian Institution's Albert Einstein Planetarium in Washington, D.C., since 1978 and has lectured widely on all aspects of astronomy. A noted astrophotographer, his work has appeared in several astronomy magazines.

DAVID H. DEVORKIN is Curator, History of Astronomy, at the National Air and Space Museum, Smithsonian Institution. His interests include the origins of modern astrophysics.

ELI DWEK is an astrophysicist in the Infrared Astrophysics Branch at NASA Goddard Space Flight Center. His areas of research include stellar evolution and the properties of dust in the interstellar medium.

MARTHA HAYNES, a radio astronomer on the faculty of Cornell University, conducts research into the large-scale structure of the universe.

FRANK J. KERR, professor emeritus at the University of Maryland, taught there for more than twenty years. He has done extensive research on the structure and motions of the Milky Way.

RICHARD LARSON is professor of astronomy at Yale University, where he specializes in the evolution of galaxies, early dynamics of star systems, and dynamical models of galaxies.

VERA C. RUBIN, an astronomer at the Department of Terrestrial Magnetism, Carnegie Institution of Washington, studies the rotational properties of stars and gas in the disks of spiral galaxies.

FRANÇOIS SCHWEIZER, one of the first astronomers to observe and study the processes involved in galactic mergers, is currently on the staff at the Department of Terrestrial Magnetism, Carnegie Institution of Washington.

JOSEPH SILK is a professor of astronomy at the University of California, Berkeley, where he lectures and writes about cosmology.

GERRIT L. VERSCHUUR, a radio astronomer who specializes in the study of interstellar hydrogen, has taught astronomy at the University of Maryland and publishes widely on the subject.

Library of Congress Cataloging in Publication Data
Galaxies/by the editors of Time-Life Books.
p.   cm.   (Voyage through the universe).
Bibliography: p.
Includes index.
ISBN 0-8094-6850-6.
ISBN 0-8094-6851-4 (lib. bdg.).
1. Galaxies. I. Time-Life Books. II. Series.
QB857.G378 1989
523.1'12—dc19   88-2281 CIP

For information on and a full description of any
of the Time-Life Books series, please call 1-800-
621-7026 or write:
Reader Information
Time-Life Customer Service
P.O. Box C-32068
Richmond, Virginia 23261-2068

Time-Life Books Inc. offers a wide range of fine
recordings, including a *Rock 'n' Roll Era* series.
For subscription information, call 1-800-621-7026
or write Time-Life Music, P.O. Box C-32068,
Richmond, Virginia 23261-2068.